Construction Rock Work Guide

Wiley Series of Practical Construction Guides

M. D. Morris, P.E., Editor

Jacob Feld

CONSTRUCTION FAILURE

William G. Rapp

CONSTRUCTION OF STRUCTURAL STEEL BUILDING FRAMES

John Philip Cook

CONSTRUCTION SEALANTS AND ADHESIVES

Ben C. Gerwick, Jr.

CONSTRUCTION OF PRESTRESSED CONCRETE STRUCTURES

S. Peter Volpe

CONSTRUCTION MANAGEMENT PRACTICE

Robert Crimmins, Reuben Samuels, and Bernard Monahan

CONSTRUCTION ROCK WORK GUIDE

Construction Rock Work Guide

Robert Crimmins, P.E.

Reuben Samuels, P.E.

Bernard Monahan, P.E.

Thomas Crimmins Contracting Co., N.Y.

WILEY-INTERSCIENCE, a Division of John Wiley & Sons, Inc.
New York · London · Sydney · Toronto

Library of Congress Cataloging in Publication Data:

Crimmins, Robert.

Construction rock work guide.

(Wiley series of practical construction guides)

1. Rock mechanics. 2. Excavation. I. Samuels, Reuben, joint author. II. Monahan, Bernard, joint author. III. Title.

TA706.C74 624'.1513 72-1809

ISBN 0-471-18753-4

Printed in the United States of America

10 9 8 7 6 5 4 3 2 1

Series Preface

The construction industry in the United States and other advanced nations continues to grow at a phenomenal rate. In the United States alone construction in the near future will exceed ninety billion dollars a year. With the population explosion and continued demand for new building of all kinds, the need will be for more professional practitioners.

In the past, before science and technology seriously affected the concepts, approaches, methods, and financing of structures, most practitioners developed their know-how by direct experience in the field. Now that the construction industry has become more complex there is a clear need for a more professional approach to new tools for learning and practice.

This series is intended to provide the construction practitioner with up-to-date guides which cover theory, design, and practice to help him approach his problems with more confidence. These books should be useful to all people working in construction: engineers, architects, specification experts, materials and equipment manufacturers, project superintendents, and all who contribute to the construction or engineering firm's success.

Although these books will offer a fuller explanation of the practical problems which face the construction industry, they will also serve the professional educator and student.

M.D. Morris, P.E.
Series Editor

Preface

The need for a construction guide that details the peculiarities of rock work has increased in recent years. Favorable or natural foundation sites have been used up, and owners are now turning to more complicated sites for development. Ecological considerations have speeded up this process; municipalities are being pressured to construct in marginal areas that often need extensive rock excavation. Sewer and drainage lines are being required in all areas, even those that involve extensive rock work. Available information on the subject of construction rock work has often been vague, and the present trend toward large, complex structures has spotlighted the necessity for tight scheduling with strict enforcement of codes and specifications.

This reference book of practical information and photographs has been prepared for those engaged in heavy construction. The book will assist the contractor in reviewing the details of specific contracts as well as the overall efficiency of his entire plant. The designer, consulting engineer, architect, or owner will find it useful in gaining a better understanding of the construction problems involved in a rock excavation, thereby simplifying the design or specifications and thus reducing the cost. Students and professors will find it useful as a reference when they approach the scheduling, management, and control aspects of a project.

To supplement our personal experiences in the field of foundations many sources have been used, and for this reason we dedicate this book to all the fine men who have entered our profession.

New York, New York
March 1972

ROBERT CRIMMINS
REUBEN SAMUELS
BERNARD MONAHAN

Contents

Construction Rock Work Guide

1

Introduction

Among construction men the term "rock" is applied to a natural aggregate of mineral particles that requires the use of explosives in excavation. The costs and physical hazards of rock excavation depend on whether the rock is intact, broken, or decomposed.

The condition of any given kind of rock can vary greatly within a single project. Rock pressures and the tendency of rock banks to fail also vary. A basic knowledge of the nature and common defects of rock beds is necessary for the efficient bidding on and performance of any rock-work project.

1.1 Geology—A Review of Basic Principles

Rocks are divided into three groups: igneous, sedimentary, and metamorphic. The inherent strengths and weaknesses of the members of these groups can be compared to the three principal construction materials—steel, concrete, and masonry.

Igneous rock such as granite was formed by the superheating under great pressures of basic elements and the slow natural cooling of this molten material after it had been introduced into or between other rocks. Slow cooling allowed large crystals, or phenocrysts, to form. Igneous rock such as basalt differs from granite in that it had been extruded from volcanoes or fissures in the earth's crust, forming streams or lakes of lava on the landscape. Owing to relatively rapid cooling, the resulting rock is aphanitic.

Sedimentary rock includes such common types as conglomerates, sandstones, shale, and limestones. The origin of the materials that make up sedimentary rocks may be earth, rock, or living matter, such as the hard portions of clams and shells. Matter is eroded in mountain areas and transported into valleys by natural phenomena, such as streams, wind, and storms. These materials, whether coarse or fine, are then cemented together by the deposition of binding materials in interstices between individual grains. Sand and gravel are gradually transformed into sandstone and conglomerate. The chief ingredients of limestone shells or coral originate in the ocean where they are solidified by the deposition of calcium carbonate, or shells, dissolved in the sea water.

Rocks formed by the consolidation of such fine-grained sediments as silt or clay are known as shales. The transformation takes place not by cementation, as in coarse-grained sediments, but by local intergrowth between adjoining grains. This intergrowth is associated with at least slight mineralogical changes. It does not take place unless the sediment is subjected to moderately high temperatures and pressures acting over long periods of time. The transition from silt and clay to shale at depth is gradual. Therefore, materials in an excavation that are in a transition stage can be of a quality that fits neither rock nor clay description.

Metamorphic rock is the result of a process of recrystallization that took place at high temperature and high pressure, which increases density and strength. Characteristics of the rock encountered depend on the nature of the material subjected to high temperature and high pressure. Limestones are transformed into marble, sandstones into quartzite, and shales into slates and schists. They are known as low-grade metamorphic rocks. As temperature and pressure increase, the rock is metamorphosed into a very hard and dense gneiss—a high-grade metamorphic rock.

Slates and schists are usually derived from shales. During the metamorphism of a shale, the percentage of micaceous minerals becomes larger and the particles become parallel. As a result, slates and schists split very readily into thin plates or slabs. The capacity to split in this manner is known as cleavage. The presence of cleavage weakens the rock mechanically so that its compressive strength in the direction of the cleavage planes is very small. Sheets of quartz parallel to the cleavage planes can increase the time and expense of drilling. In gneiss the percentage of flaky constituents is much smaller than in schist. Therefore, cleavage is erratic and strength can be high. The mineral composition of gneiss is generally similar to that of granite.

Every rock has mechanical defects, or fractures. Simple fractures along which no noticeable displacement has occurred are called joints. A joint can be opened or closed and may not be visible. There is no resistance along a joint surface. In quarries the joint spacing determines the largest size of blocks that can be obtained. Joints in igneous rocks are associated with the volume contraction associated with cooling. Many deformed rocks have joints that are due to failure by tension. In numerous rocks the joints are either not continuous or so irregular that blocks located between them are intimately interlocked. These joints break the continuity of the rock and reduce the average strength of the jointed mass to a small fraction of its intact strength.

Complex fractures of large extent along which relative displacement has occurred are called faults. The rock adjacent to a fault is usually crushed. The fault may or may not occur along a bedding plane, or the surface separating layers of sedimentary rock, which is a plane of weakness.

In contrast to the processes that lead to jointing, faults and folds are developed by high compressive stresses that have been caused by natural adjustments in the earth's crust. The existence of these weaknesses are commonplace and must be assumed to exist if sufficient information to prove their nonexistence is lacking. Faults and folds require special construction procedures and should be approached with caution.

Common types of faults are the following:

1. Normal fault (see Fig. 1.1)—hanging wall appears to have moved downward.
2. Reverse fault—hanging wall appears to have moved upward.
3. Strike-slip fault—horizontal movement with slipping parallel to the strike of the fault.

In folds, the convex part is called an anticline, and the concave part is called a syncline. The relative position of the sides of the fold determine the type of fold, whether symmetrical, asymmetrical, overturned, or recumbent.

Recognition of the type of fault or fold is necessary in order to determine the type of material that will be encountered during construction.

Fig. 1.2, 1.3, 1.4, 1.5a, 1.5b, 1.6, 1.7 and 1.8 illustrate the scope of rock excavations.

1.2 History of Rock Excavation

The history of rock excavation is written in the hearts of men who have withstood and overcome the obstacles of nature. The earliest types of rock excavation were man's attempts to protect himself by fashioning a shelter in the side of a rock cliff. Recorded history indicates that the Romans used "advanced" methods to split rock formations for both road building and

Fig. 1.1 Structures are often built on the side of a rock slope. Natural bedrock formations are rarely level.

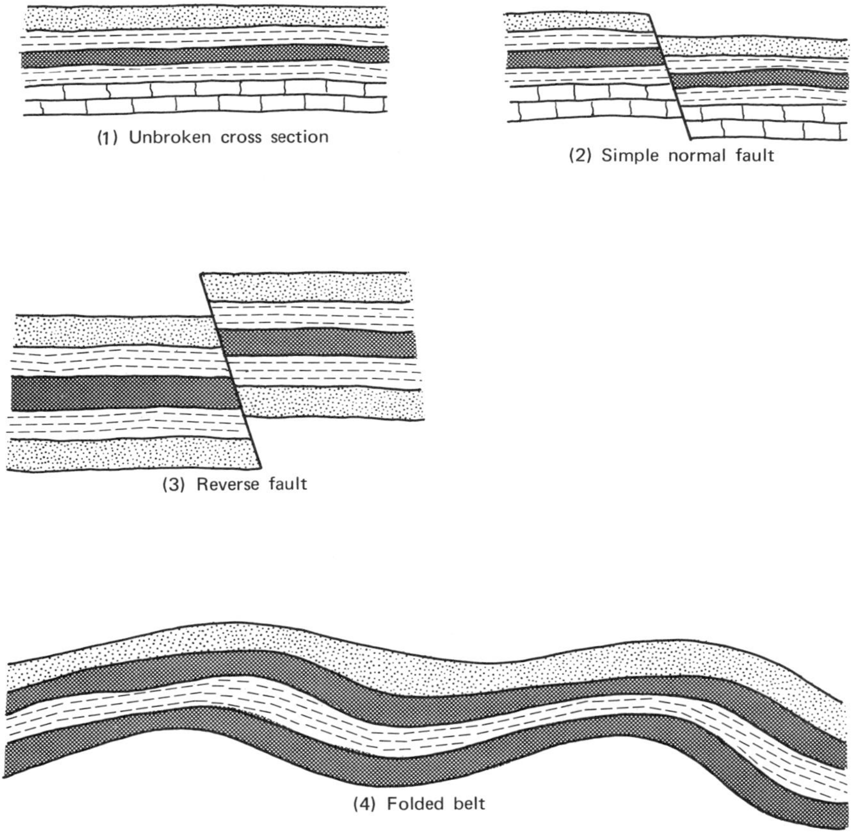

Fig. 1.2 Natural rock formations usually fall into one of the above classifications.

quarrying. Ancient Egyptians built the pyramids by splitting dense rock formations into massive building blocks, and then piling them together to form the great pyramids.

Early foundations were constructed almost entirely by brute force that required great endurance by the skilled men who did the work. Today machinery is often substituted for manpower; but it is still an endeavor requiring workers of skill and endurance.

Figure 1.9 demonstrates the appearance of a rock formation after being cut by modern tools and equipment to form a depression for a building footing.

Tunnels have been in existence for many thousands of years. The earliest civilizations built tunnels to mine precious metals. These mining tunnels

Fig. 1.3 Limestone beds can be upended during geological disturbances. Niagara Falls, 1969.

were bored into the rock or overburden, and early experience dictated the settings of supports so that the men could work in relative safety.

Trenches were used by early civilizations to cover conduits supplying gravity-fed water to a city.

Man continually refined hand methods of excavating rock, and in the early 1700s first used explosives for excavation purposes. In the United States the first recorded application of explosives for rock excavation was in 1780 at the Simsbury Copper Mine in Connecticut. Explosives were utilized to help excavate cells of the main shaft. Blasting does not remove material; rather, it changes the form of material to simplify removal.

Until this period the art of tunneling was limited to mines for precious

Fig. 1.4 Sedimentary rocks, such as shales, are transformed into schists by the natural process of metamorphism.

Fig. 1.5*a* Overbreakage is common in highly disturbed formations.

Fig. 1.5*a*, *b* Joints, faults, and folds are common in geological areas that have been subjected to tectonic disturbances.

Fig. 1.6 Joint systems exist in the majority of natural rock formations. New York Thruway—Kingston area, 1969.

metals. Hand methods were applied to dislodge the minerals. The discovery of the many uses of black powder as an explosive in tunneling was to alter the future of rock construction.

The first major United States job on which dynamite was successfully used was the driving of the Musconneteong tunnel on the Lehigh Valley Railroad. This tunnel, a little less than a mile long, was started in the fall of 1872. At times during its construction, as much as 17,000 lb of dynamite were utilized a month.

In 1885 an aqueduct tunnel was driven from Ceoton to New York City. Thereafter, blasting was determined to be a safe and reasonable method for solving engineering problems in the design and construction of tunnels and foundations (see Fig. 1.10).

Fig. 1.7 Excavation for the Catskill Aqueduct, 1922, passed through a multitude of rock formations that were classified in each of the three main categories: igneous, metamorphic, and sedimentary.

Fig. 1.8 Bowery Savings Bank, 1921. Highly disturbed strata may require the use of specialized procedures to prevent bedding planes from developing into failure planes. The cutting of a vertical slot in the formation and its subsequent filling with concrete serves to transfer the footing loads to a more secure base. This underpinning procedure in rock is still used when excavating adjacent to buildings founded on inclined slopes.

Fig. 1.9 Prehistoric man may have cut depressions in rock in search of precious metals or shelter, similar to cavern cuts used to support building foundations.

1.2.1 *Introduction of High-Power Explosives*

Gelatin dynamite was used in quantity to drive the early aqueduct tunnels. Our present-day "blasting gelatin" is modeled after Alfred Nobel's discovery in 1875: nitroglycerin dissolved in collodion cotton resulted in a gelatinous mass more powerful than dynamite.

Truly the history of rock work cannot be separated from the history of explosives. Black powder, a mixture of 75 percent saltpeter, 15 percent charcoal, and 10 percent sulphur, was used as a firearm propellant until the 1600s, when experiments were initiated to harness this energy for larger jobs. In England in 1831 William Bickford invented the safety fuse, which provided the first reasonably safe method to ignite black powder.

In 1802 E. duPont introduced the substitution of Chile saltpeter for potassium nitrate, materially lowering the cost of his product. This invention helped to increase the consumption of black powder from 25 million lb in 1860 to 98 million lb in 1900!

Historically, men have sought to overcome the obstacles of a natural rock formation by persistence in the use of hand implements and, later, by the use of explosives. In reviewing our progress in the last century one can only surmise that the future will surely accelerate technical knowledge to the point where it will challenge tomorrow's engineer to an ever widening horizon (see Fig. 1.11). Surely methods will be commercialized whereby bedrock will be liquified under pressure to form molten lava, which may be removed by bucket or pump. Possibly saws will be perfected that can simply cut the bedrock into handy truck-sized chunks.

Nuclear fission (as in the atomic bomb) and nuclear fusion (as in the hydrogen bomb) eventually will be harnessed to use as commercial explosives.

Fig. 1.12 and 1.13 (W13, W14, W15, W16) illustrate the complexity of foundation construction.

1.3 Job Classifications—Shafts Footings, Tunnels, and Trenches

Job classifications that may require rock excavation for successful completion are shafts, building foundations, tunnels, and trenches. Shafts are usually constructed at tunnels to provide access for ventilation and egress for waste materials. The earth overburden must be evaluated for probable pressures. If it is saturated, steel sheeting and extensive dewatering may be required to house the shaft prior to the start of rock excavation.

A common method of investigating a site is to dig a test hole, usually with a small backhoe, or to drive a well point, similar to that used for pri-

Fig. 1.10 Rock drills and dynamite were quickly adapted to facilitate rock excavations for building foundations, as well as tunnels, as seen in this early (1921) building foundation.

Fig. 1.11 Early construction projects were hazardous to workers and on-lookers alike.

Fig. 1.11 Hydraulic backhoes are the most recent improvement in the rock excavation field. They are dependable and adaptable to varying conditions.

Fig. 11.1 Mobile cranes and shovels have for the most part displaced the earlier derrick.

Fig. 1.12 Brick columns are no longer used because of their relatively low compressive strength; however, line drilling, as shown above, has remained as a sound and efficient means of predetermining the extent of removal. Line holes are drilled on the basis of 2.5 holes/lin ft. Depths of 20 ft are standard.

Fig. 1.13 Light charges can be used to fracture bedrock formations, even in congested areas. The photo above was taken in the Grand Central terminal tunnel (New York City) in the early 1930s. Rope mats were used to cover charges and hold down fragments. More recently, construction projects in this tunnel were completed without the use of explosives because of the need to maintain commuter service.

vate wells, and then to make periodic checks on the water level in the test hole or the test-well point. These results, when correlated with information supplied by private citizens, public officials, and other contractors, can lead to an excellent prediction of probable soil pressures and sheeting requirements.

After the sheeting has been secured with walers, braces, or tie backs, the contractor is then faced with the problem of excavating the rock within the shaft to the desired depth (Fig. 1.14). If the opening is small, less than 50 ft^2, then the contractor can elect to use air hammers to remove the bedrock; but if the shaft contains sizeable quantities of rock that must be excavated, then blasting techniques are usually more efficient.

A common type of shaft that occurs frequently in the construction of commercial buildings is the "column footing." These footings may vary in size from 9 to 200 ft^2, depending on the allowable bearing of the rock and column loads. A common size is 20 ft^2 on a design bearing of 40 tons/ft^2, or a column load of 1600 kips.

Column footings are usually excavated with air tools and hand labor, using methods developed by our forefathers at the turn of the century. Explosives are not used because the depth of the footing is determined by the hardness of the rock. When the rock becomes hard enough to require explosives, the excavation is halted and the footing is poured.

The method of debris removal in a small footing shaft is usually well-wheel-and-bucket. The workman fills the bucket, tugs on the rope, and the bucket of debris is hoisted to the top by the laborer stationed at the surface. Larger shafts are excavated using a crane and tip bucket that may be loaded with a cubic yard of rock debris before being hauled to the surface and loaded onto a truck, which then dumps it in a disposal area.

Building foundations often require the removal of thousands of cubic yards of earth and bedrock. Earth and soft rock usually can be handled by shovels, bulldozers, and other mechanical equipment.

Rock must be blasted prior to being excavated. If the building is located in a crowded metropolitan area, special safety precautions must be followed in order to control vibrations and extent of flying debris. Advances in blasting technology have enabled the foundation contractor to follow safe procedures while removing rock at a profit.

Foundation depths are increasing because of the rising land values and demand for storage, parking, and mechanical space in our modern buildings. Over the last three generations building costs have soared, while foundation costs have risen only moderately because of improved methods. This has resulted in an increasing need for competent contractors who are knowledgeable in rock work.

Fig. 1.14*a* Early construction projects featured cross lot bracing to support earth banks.

Fig. 1.14*a*, *b* Sheeting and shoring of overburden must be secured prior to making the initial rock cut. Blasting of rock strata further loosens any carelessly installed braces. Vertical tiebacks must be secured against downward creep because any vertical displacement is accompanied by a similar horizontal movement and rotation.

2

Prebid Factors

2.1 Project Evaluation

Construction jobs of all sizes and shapes must be well thoughtout before drawings are finalized and before bids are submitted.

The design engineer must inform the owner of existing limitations on the development of certain concepts. Often a novice real estate developer purchases a plot with the idea that the area can be built on at average foundation prices. If the plot is underlain by bedrock, then the cost of removal is usually two to three times the expense of earth excavation. (Fig. 2.1)

It is common practice for professional real estate developers to consult competent contractors for budget estimates of probable costs. To provide a cost estimate for a foundation or structure in bedrock, the contractor should consider the following: (See Fig. 2.2, 2.3, 2.4, 2.5.)

1. Size of plot.
2. Depth of excavation in earth.
3. Depth of excavation in rock.
4. Water level.
5. Nearness of existing structures.
6. Sheeting and shoring requirements.

The size of plot determines the probable progress rate. A large plot could provide access for trucks to several shovels; a shaft could require handloading all materials into buckets, which would then be hoisted to the surface and loaded in trucks for disposal. In a city-core area a large quantity of rock can be blasted and removed for less than \$15/yd^3; a pier

Fig. 2.1 Rock-excavation work requires careful analysis before submitting a bid. Line drilling and shoring adjacent to existing structures must be included. Closely spaced holes are required to minimize shock vibrations under adjacent buildings. Note that the blasting operation has removed only the rock inside the line holes.

Fig. 2.2 In crowded metropolitan areas, blasting operations must be carefully supervised. Bidders should carefully evaluate the traffic pattern in the area of a given project.

Fig. 2.3 Large plots may lend themselves to rapid excavation with two or more shovels, as shown in this photo, taken in 1938, at the site of the Metropolitan Life Insurance Company in Manhattan. The existence of old building foundations may increase the cost of earth excavation to that of rock excavation. Bidders are advised to investigate the site before submitting a bid.

Fig. 2.4 Metropolitan Life Insurance building, 1938, on East 25th Street and Madison Avenue in New York City. Timber ramps are often a necessity and should be allowed for in bidding the work. Heavy-duty construction is required with loads of 75 tons (not uncommon). Costs in the 1969 market are in the $70/lin ft range.

Fig. 2.5 Foundation contractors are required to design and install temporary ramps that are capable of supporting construction loads, such as the illustrated concrete truck. These are all weather ramps that usually remain to be used by subsequent contractors, such as the stell erecta with his heavy-duty cranes.

hole, that is, a footing or shaft, might require hand labor, which often increases the cost to well over $75/yd^3$.

If loose sand extends over the whole area, earth overburden may be removed easily for less than $5/yd^3$, including trucking; but if the overburden is interwoven with existing foundations, blasting may be required, thus increasing the overall cost. Hardpan, a conglomerate of gravel cemented with clay, can approach bedrock in difficulty of removal.

Rock should be evaluated for hardness and type before estimating excavation costs. Igneous rock, such as granite or basalt, can shatter easily and require less explosives than softer rocks. A hard igneous rock transmits shock waves farther than softer or less-dense materials. This is a consideration in excavating near existing structures. Sedimentary rock, such as limestone and shale, can be ripped or lifted by horizontal blasting, thus decreasing the overall costs. Metamorphic rock, such as schist (low-grade or soft rock) and gneiss (high-grade metamorphic rock), is variable in its cost of removal. Generally the softer rocks are less expensive because of their ability to be ripped and the smaller bulking factor.

The presence of water materially increases the cost of excavation. Sump trenches must be excavated first to relieve the flow. Pumps must operate, possibly continuously, to prevent flooding of the construction site. Metamorphic rocks have an affinity for water and soften after reaching saturation. The consequence of this natural law may be far reaching in that footings, if to be founded on firm bearings of, for example, 40 tons/ft^2 (as described in the New York City building code), may have to be excavated much deeper than the original minimum anticipated. The common occurrence of this variability in rock quantity is a factor that has led many foundation contractors to an early bankruptcy.

The nearness of such structures as buildings, water mains, gas mains, and public streets may determine the construction procedures and thus the cost of work. An adjacent masonry foundation wall can be disturbed by blasting vibrations or excavating equipment. Water and gas mains can be ruptured by severe vibrations or by undermining of supports. Streets can crack or settle as a result of adjacent excavation.

Vertical banks must be sheeted, adjacent buildings must be shored or underpinned, and adjacent yards or property must be protected and maintained.

A proper prebid evaluation must take into consideration all these items and many more that are less obvious.

2.2 Contingencies Excluded from Bid

Contractors normally exclude those contingencies requiring an allowance of monies that would overload their bid. Common exclusions or items for which separate sums are usually provided are the following: (Fig. 2.6, W24)

1. Depth of footings—usually a unit price is provided for extra depths (0 to 5 ft, 5 to 10 ft, and 10 to 20 ft).
2. Pumping—can be limited by intervals and by shifts. That is, a contractor's bid may specify that pumping has been figured on a normal work day basis for an interval of 6 months. Any additional pumping requirements are then paid for by the owner.
3. Maintenance of banks can be a minimum under certain conditions, such as a large site, and ability to slope cuts while steel sheeting may be required if streets are to be maintained and water levels are to be interrupted. For this reason sheeting is either excluded or the extent is clearly specified. If additional sheeting is required, it may then be paid for on a unit price basis.

Controversies can be avoided if contractors and owners demand clearly outlined contracts. Borings and experience can combine to warn competent contractors to stand aside while incompetent fly-by-nights take unnecessary gambles that often lead either to bankruptcy, incompleted contracts, or both. Subsurface contingencies should be outlined in the bid proposal and monies should be provided; appropriate exclusions should result in a savings to the owner.

2.3 Off-Site Factors

Prior to submitting a bid involving the removal of large quantities of excess material a thorough investigation must be made into the availability of dumps and their convenience to the job site. In city areas permits must be obtained to cross the sidewalk, and the traffic department must be consulted concerning the interruption of vehicular flow. In New York City, areas suitable for disposal of excavated rock are particularly scarce. It is not unusual for a truck to have a 2-hour cycle, that is, 1 hour from job to dump and 1 hour to return.

In metropolitan areas dumping charges can affect the cost of excavation. Hidden costs, such as maintenance of the dump, can also be a factor. It

Fig. 2.6 Existing utilities, such as water mains and storm drains, can often lead to a contractor's loss. The flooding of the site above by a ruptured main was a contingency that was not anticipated.

is common practice for the owner of a dump to charge for disposal, and to require a bulldozer ($250/day) at the dump for maintenance.

Traffic, bridge, and tunnel departments should be consulted about probable delays. Often the hauling schedule is determined by off-site conditions over which the contractor has no control. It is not uncommon for trucking to be restricted to either evening hours or nonpeak hours (10 to 2) on highly congested thoroughfares.

The bidder should be aware of hidden pitfalls that only years of experience can expose (Fig. 2.7a, 2.7b). A good practice before bidding a project is to cover your bid prices by receiving bids from subcontractors familiar with the area and the conditions. Then, after the contract is awarded, you can decide which operations are particularly attractive, and perform these with your own forces.

2.4 *Rock Failure*

Rock work is the removal of sections of geologic formations that interfere with the proposed project. This interruption of nature's domain redistributes the stresses formerly transmitted through the excavated formation.

The removal of the toe of inclined strata can cause the strata to overcome any inherent frictional resistance and slide into the excavation. The raising of a groundwater table, as in a dam project, can cause inclined strata to overpower frictional resistance by floating and thus cause a landslide. Blasting vibrations can destroy any natural friction that might have held inclined strata from sliding.

Bidding a project requires the careful evaluation of the effects of the proposed construction. Specifications usually hold the contractor responsible for safeguarding the surrounding areas. See Fig. 2.8, 2.9. On various occasions, courts have held the owner, designer, or contractor responsible for failures that could have been predicted by a careful evaluation of the geologic conditions and the unknown variables.

If natural strata are inclined and there is any question about their stability, then provision for expert design must be made. Inclined strata can be secured by tie rods grouted into a base stratum, or they can be removed entirely, depending on the economy and feasibility of individual alternates.

Borings do not always show the required information. Expert geologic investigations may be required before bidding. A contractor should not bid any project that seems to be lacking in subsurface data unless he has acquired the experience necessary to predict probable geologic behavior in the specific area.

Fig. 2.7*a* Hand rock removal—progress depends on the quality of the labor force as well as the hardness of the rock.

Fig. 2.7*a*, *b* Tip buckets are used to transport material to waiting trucks, but actual disposal at private or public dumps must be considered before the work is bid.

Fig. 2.8 Newark City Railway, 1931. Vertical rock cuts always present some danger of sliding. Careful observation and the removal by scaling of loose and disjointed blocks minimize these dangers.

Fig. 2.9 The excavation of inclined rock strata can often lead to a disruption of static stresses. Care and supports must be provided to preclude disaster.

2.5 *Insurance Requirements*

Insurance protection is a recognized prerequisite for bidding any project (see Fig. 2.10). Rock work, because of its inherent dangers due to flying debris and falling objects, has the highest premiums in the industry. These premiums are then adjusted to conform with the contractor's experience or safety rating. It is standard practice for specifications to outline the insurance coverage that minimizes the liability of the owner and the design engineer. The contractor then must decide what additional policies or riders are required to safeguard his investment and his business.

Insurance premiums commonly depend on the contractor's labor payroll, and the range of rates may be in the 20 to 40 percent of payroll area. An improvement in a contractor's experience rating thus lowering his premiums is often the difference between a profitable undertaking and a disastrous project. The contractor should advise his insurance agents of his intent to bid a certain project, and request them to furnish a firm proposal for insurance coverage.

Specifications requiring that a qualified safety engineer be on the contractor's staff usually lead to lower insurance premiums and a safer job. An excellent job, safetywise, usually leads to a profitable future job. Safe jobs are usually on schedule and efficient.

Liability Insurance

General Liability Insurance	
Bodily Injury	\$1,000,000/person \$3,000,000/occurrence \$3,000,000 aggregate products
Property Damage	\$1,000,000/accident \$1,000,000 aggregate contractual

The policy shall be endorsed to eliminate the exclusions for explosions, collapse, and injury to and destruction of underground utilities and for property in the care, custody, and control of the insured.

Workmen's Compensation is required.

Fig. 2.10 Insurance requirements for foundation contractors.

2.6 *Bid Forms*

Standard bid forms must be set up and adhered to on all projects in which a contractor participates. A major cause of contractor losses is the lack of a realistic bid form that accurately reflects the true costs to be expected. Rock work in metropolitan areas usually requires inclusion of many items to be performed while the excavation is in progress. Rock work on large dams or freeway projects may often include setting up work camps and supplying necessary services to the workers and their families.

The first step in preparing a bid is to determine the quantities required for the completion of the project. These quantities must be grouped together into descriptive items that indicate the anticipated cost of performing the work. For example, "General Excavation of Rock" would indicate the work to be performed with the least difficulty, whereas "Footing Excavation of Rock" would probably be assigned a unit of much higher value in the estimate (see Fig. 2.11).

Trench or pit rock would be assigned an estimated cost at an intermediate point on this sliding scale. Many projects are let on the basis of quantities estimated by the designers. The contractor must be sure his estimated quantities are computed in a similar fashion. Pay quantities for rock excavation may often exclude the overbreak that actually occurs. Specifications and bid instructions usually outline the method of payment, which the contractor must then adhere to for bidding the project. If the method of computing quantities is not clear, then the bidder should state the method to be used in the bid proposal for determining payments for contract items and extra work.

Estimated material costs can then be projected by using the pay quantity, plus any anticipated overrun, to determine the probable material cost. Concrete is the simplest example: a supplier charges \$20 to \$30/yd^3 of concrete delivered, the bidder adds a percentage (5 to 15%) to cover waste or overpour, and the unit is then multiplied by the quantity and the total inserted in the estimated material cost column. Rock excavation may entail dumping charges or the purchase of materials, such as rock bolts, necessary to the performance of the contract. Subcontractors' bid costs can also be included in this section.

Estimated equipment rental costs should not exceed the purchase price of the item unless the contractor feels that maintenance or depreciation equalizes the difference. A contractor's estimated equipment costs should reflect projected costs rather than profits that would be inherent in a supplier's handbook of rentals. Equipment should therefore be divided into items that are to be rented from suppliers on an as-needed basis and tools

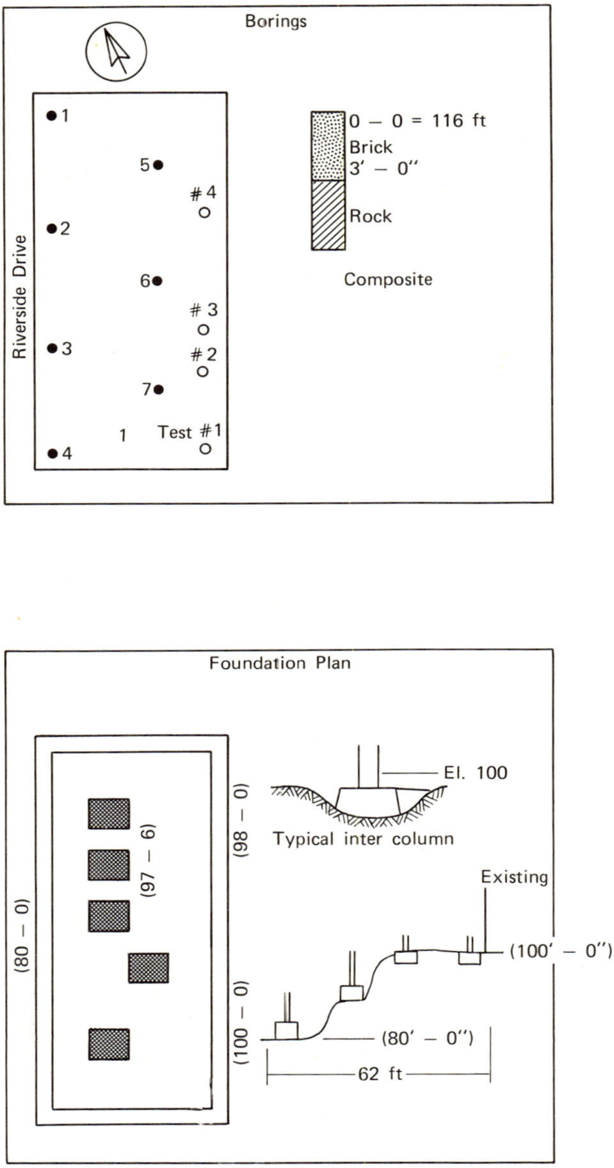

(a)

Fig. 2.11 Contract plans, to be used in bidding a foundation project, often vary in nature and scope.

Summary	Quantity	Unit	Total
Construction fence	310 l.f.	$7	$2,170
Clear & grub	18,000 ft^2	0.10	1,800
Reinforced chain link fence	310 l.f.	2	620
Reinforced sidewalk	2,000 ft^2	0.30	600
General rock excavation (sub. 6.25)	7,500 yd^3	20	150,000
Rock excavation—column footings	30 yd^3	100	3,000
Pits	500 yd^3	70	35,000
Mechanical trenches	180 yd^3	100	18,000
Level bottoms—column footings	400 ft^2	10	4,000
Wall footings	500 l.f.	5	2,500
Line drilling—Perimeter	76,000 ft^2	3	22,800
Pits	2,900 ft^2	2	5,800
Trenches	3,800 ft^2	2	7,600
Footings	900 ft^2	2	1,800
Grading	12,000 ft^2	0.20	2,400
Pumping	L.S.		1,000
Ramps & runways	L.S.		5,000
Backfill	800 yd^3	10	8,000
Rock seal (concrete)	50 yd^3	100	500
Porous pipe (4 in.)	300 l.f.	5	1,500
Porous fill	5 yd^3	10	500
Plumbing	L.S.		7,500
Electrical	L.S		1,000
			283,090
G.C.			21,910
			305,000
P.C.			35,000
			$340,000

Fig. 2.11*a* Maxwell Hall addition, Presbyterian Hospital, April 1969. (Thomas Crimmins Construction Company.)

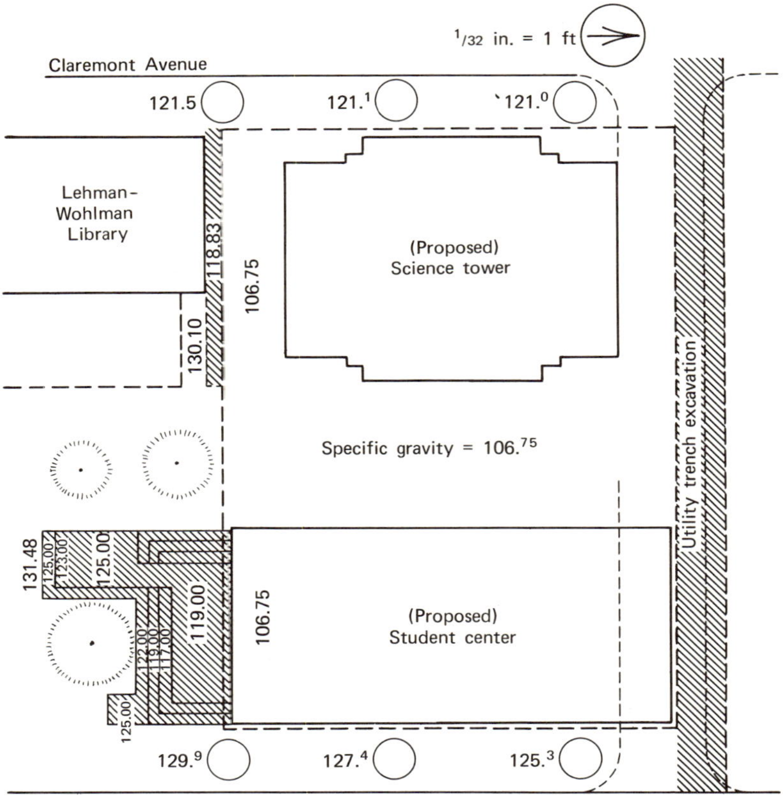

Fig. 2.11*b* Barnard College, 1966.

that are either owned or to be purchased for the job. A company-owned bulldozer to be used on the project is charged by subtracting the probable value of the machine after project completion from its value at the time of the project's inceptícn. The result divided by the job duration gives a fair projection of equipment rental rates.

Estimated costs for bidding purposes are determined by experience, and probable labor costs for rock work are most assuredly the hardest items to estimate. Labor costs vary with the season of the year, the quality of the workers, the type and characteristics of the rock, the size of the job, and any other item that could be imagined to effect the job. The best procedure for estimating labor costs in rock work is to determine from experience the probable range of costs between an efficient job and an inefficient job. This range is then narrowed down to a probable cost for the projected

Barnard College—Student Center—1966

Summary

Construction fence	840 lin ft @ \$ 9.00 =	\$7,560
Remove trees, walls, etc.	allow	\$5,000
General earth excavation	11,200 yd³ @ \$ 2.00 =	\$22,400
General rock excavation	11,850 yd³ @ \$ 7.50 =	\$88,875
Trench and pit earth	200 yd³ @ \$10.00 =	\$2,000
Trench, pit, and footing rock	762 yd³ @ \$50.00 =	\$38,100
Sheeting	2130 ft² @ \$ 3.00 =	\$6,390
Backfill	235 yd³ @ \$ 5.00 =	\$1,175
Grading	30,000 ft² @ \$ .10 =	\$3,000
Line drilling	8,170 ft² @ \$ 1.00 =	\$8,170
General items (doors, lights, etc.)	allow	\$6,330
		\$189,000
	General conditions 6%	\$11,340
		\$200,340
	Profit and overhead	\$20,060
		\$220,400

Fig. 2.11*c* Barnard College—Student Center.

job. Labor cost estimates must be based on the past cost records, including rainy and inefficient days as well as high-production days. It is easy for an estimator to confuse desired costs with actual costs. Superintendents and foremen are notorious for underestimating actual costs. We all have a tendency to remember only the days of high production, and to base a bid on these experiences can often lead to losses for a contractor.

Actual costs of previous projects similar to the prospective bid being studied are helpful in determining estimated costs. Past-project costs are prepared by analyzing account records and daily job reports. The account records give the total monies spent for (a) labor, (b) insurance, (c) materials, and (d) equipment. The daily reports should help breakdown "labor" into cost categories that accurately reflect total labor costs of shop stewards, inclement weather, and items of delay. This reconstruction of costs should then be revised to indicate labor increases.

3

Rock Work Equipment

The cost of completing a typical rock foundation has not risen as sharply as has the cost of living over the past 30 years. The reason for this comparatively slow rise in costs is attributed to advances in the development of tools that have increased productivity at a rate enabling the cost of work performed to remain relatively stable.

Equipment manufacturers spend considerable sums on research to help develop products to suit a contractor's needs. This research is often complicated by practical considerations, such as union agreements that require additional personnel for operation. Flexibility in the design of rock-work equipment has been the key to success for many manufacturers. A strong follow-up communicating design weaknesses to engineering staffs helps to assure a manufacturer that the tool he is developing will remain competitive. A manufacturer who thinks research is a wasteful drain on profits is in for a rude awakening! Research must be continuous to be effective.

A contractor must also experiment with various types of equipment to assure that his firm has lower costs than his competitors'.

A joint effort by both manufacturer and contractor often leads to costs that outstrip the competition. Manufacturers are eager to provide consultations on equipment problems. They supply specialists, often without charge, to help correct equipment operation deficiencies.

Contractors and design engineers will do well to call on these specialists before making decisions about specification requirements.

3.1 Site Investigation

Construction projects that include rock foundations also feature specialized equipment for investigating the rock and, later, for removing portions of it.

Rock investigating equipment may consist of boring rigs that core and recover bedrock samples or of geophysical apparatus that performs a generalized survey outlining the subsurface formation.

Rock boring rigs may be self-contained units mounted on a truck or jeep, or they may be simply an A-frame apparatus set up on skids at a given location. These rigs can be set up for either percussion (Fig. 3.1) or rotary drilling (Fig. 3.2).

If the drill hole must be advanced through soft rock to shallow depths, percussion drilling may be advantageous. In this method, a heavy drilling bit is alternately raised and dropped in such a manner that it grinds the

Fig. 3.1*a* Standard penetration test while making Gow test borings on the site of the Mid-America home-office building in Chicago for Prudential Insurance Company. Sixteen Gow test borings were made through clay, hardpan, and rock to depths of 135 ft. (Raymond concrete pile.)

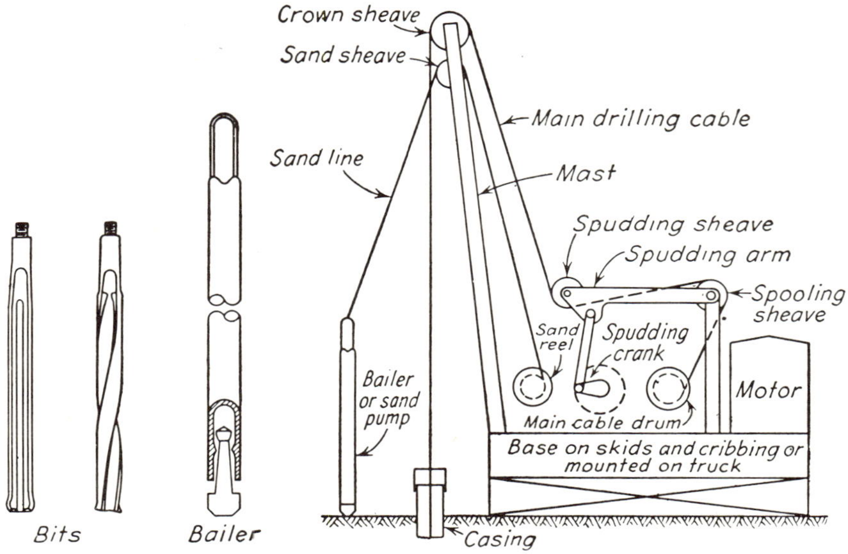

(b)

Fig. 3.1*b* Percussion drilling utilizes common soil sampling equipment and may be advantageous for shallow penetrations.

underlying material to the consistency of sand or silt. If possible, the borehole is kept dry, except for a small amount of water that forms a slurry with the material ground up by the bit. When accumulation of slurry interferes with drilling, drilling tools are removed from the hole and the slurry is cleaned out with a bailer. The hole can be cased if the formation does not stand without collapse.

Rotary drilling is the most rapid method for penetrating highly resistant materials. In the rotary drilling method, a rapidly rotating bit cuts or grinds the material at the bottom of the hole into small particles. This material is removed by circulating water or drilling fluid. If the hole tends to collapse, a thixotropic* slurry or fluid, consisting of volcanic clay in water, is used. This slurry, known as drilling mud, coats and supports the sides of the drill hole and seals off permeable strata. If drilling is interrupted, the "mud" gels and holds the rock cuttings in suspension so that they do not accumulate and jam the bit in the bottom of the hole.

* Thixotropic—describing a gel's property of becoming fluid when disturbed and again setting to a gel when allowed to stand.

Fig. 3.2a Standard Gow boring rig with drill-head unit drilling rock. Box in foreground contains rock cores recovered by the drilling unit. (Raymond concrete pile.)

3.1 *Coring*

Short, intact samples of rock can be obtained during percussion drilling by means of a special core barrel. However, the necessity for taking a core greatly slows this drilling process. Therefore, percussion drilling is rarely used in sampling operations for exploratory purposes.

In rotary drilling, the bit can be replaced by a core drill bit (Fig. 3.3) attached to the lower end of a core barrel (Fig. 3.4). The drill makes an annular cut in the rock, leaving a central core that enters and is retained in the core barrel. The cutting element may consist of diamonds, chilled

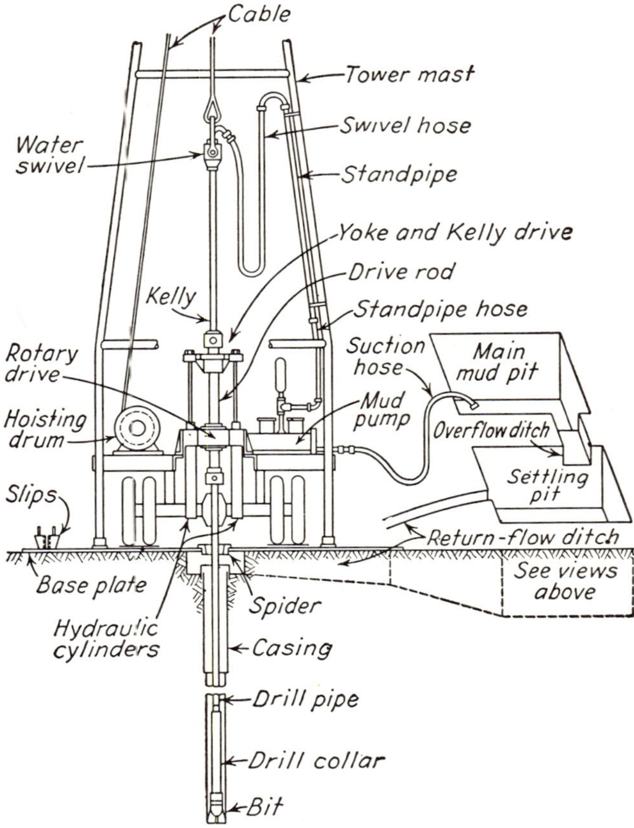

(b)

Fig. 3.2*b* Rotary drilling rig (after M. J. Hvorslev). Rotary drilling is rapid and efficient in penetrating rock strata. Oil exploration teams use king-sized rigs to penetrate great depths.

shot, tungsten carbide inserts, or steel cutters. Core drilling does not greatly slow the drilling progress and is widely used for sampling.

Cutting from the annular channel is commonly washed away with the drilling fluid, but in some materials action of the fluid alters the character of the material. Under these circumstances, cuttings can be removed with an air jet or a ring-shaped auger located directly above the cutting tool.

Core barrels may be single or double-tubs. Samples taken in a single-tube barrel are likely to be disturbed by torsion, swelling, and contamination with the drilling fluid. The double-tube barrel is designed to protect the core against action of the circulating fluid.

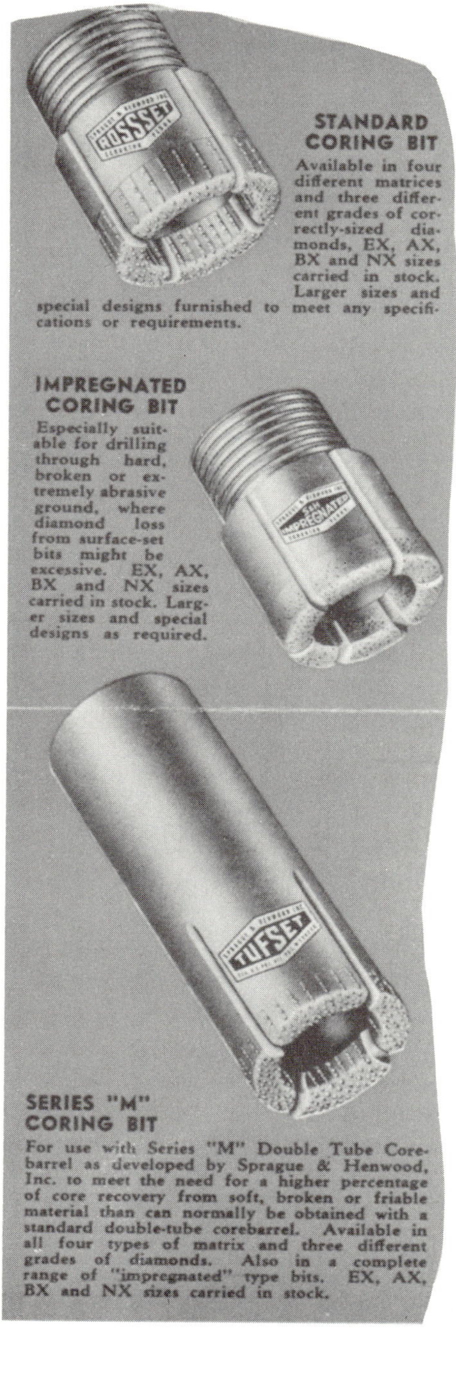

Fig. 3.3 Coring bits.

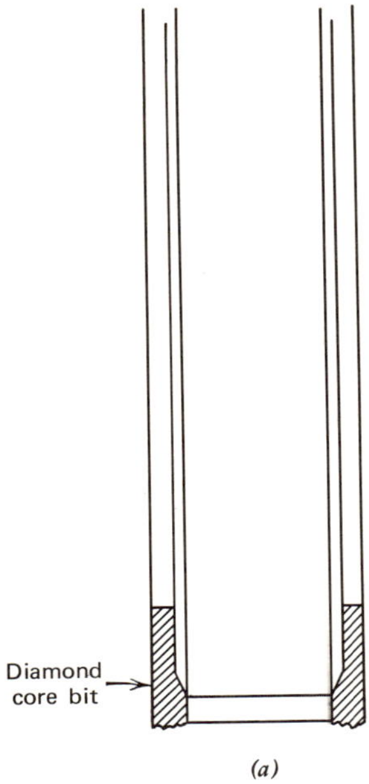

Fig. 3.4a Core barrels. Double-tube core barrels are recommended because they give greater core recovery than single-tube barrels. Double-tube core barrels are made from extremely thin tubing, and therefore special wrenches should be used. Diamond core bits, used with reasonable care, give better cores than carbide bits.

Drilling for core recovery requires a good mechanic who is interested in the work. Attention to the work is most important, for carelessness leads to poor cores and damaged bits. A bit run at high speed after the diamonds have flaked off quickly disintergrates through the abrasive action of the loose stones. Running a bit dry (i.e., when water is blocked) causes the bit to overheat and the metal setting for the diamonds to fuse into a hopeless mass.

The key factor in subsurface exploration is knowing the limitations of the equipment and the operators. Quality controls must be maintained to guarantee reasonable data on which designs may be based. These data are entered in a boring log (Fig. 3.5) for easy reference.

3.2 *Building Codes—Determinants of Construction Equipment*

Land owners who wish to develop their sites by means of construction are confronted with a building code that usually outlines acceptable criteria on which a design may be developed.

Fig. 3.4*b* A representative sample is taken from the sample spoon and placed in the sample bottle (upper photo). Sample data are recorded on bottle (lower photo) for review at nearest office.

Thomas Crimmins Contracting Company

Record of Borehole No. 6

For: BPM Construction Company
At: Ford St., Brooklyn, N.Y.
Started: 9/26/69 Finished: 10/2/69
Diameter: 1⅜″
Drill Foreman: Frank Malker

Strata
Soft schist
Medium schist
Hard gneiss

Fig. 3.5 Drilling log. Permanent bore-hole log should include any unusual features such as large seams, openings, and water tables, as well as change of strata and rock identification.

Before preliminary foundation designs can be submitted to planning or building departments, borings must be taken. New York City* requires one boring for every 2500 ft^2 of building area that is to be founded on conventional footings. If rock is exposed, the boring program can be minimized to determine only the characteristics needed by the designer. These requirements include the determination of defects or the inclination of bedding planes that can affect the stability of the foundation. Where bedrock is covered with a mantle of earth, boring must be cored a minimum of 5 ft into rock to obtain a recovery of at least 35-percent core.

Code requirements often determine the type of equipment to be used in the site investigation. The regulation of this preliminary phase, as well as the actual construction, usually leads to a practical approach to safe completion. An area that does not have a comprehensive building code often invites speculative approaches to design and construction. Experimental designs or construction practices that fail to exercise cautionary controls often lead to disasters.

Building codes and city ordinances determine the size and type of equipment that a contractor can efficiently use on a given project, such as using heavy-duty trucks on city thoroughfares. Open trucks are prohibited because of their tendency for spillage.

* See the Appendix.

3.3 Geophysical Equipment and Methods

To determine the suitability of a site, geophysical methods are used to find such rock characteristics as electrical resistivity, gravitational or magnetic field, or conductivity of sound or shock waves. Changes in these characteristics show changes in rock profiles.

The two most popular geophysical methods are the seismic method and the resistivity method.

1. The seismic method (Fig. 3.6) uses small charges set off at the surface to produce sound or shock waves that are picked up and timed at distance intervals from the point of explosion. In some cases, a shock wave is set up with a hammer blow or sharp percussion effect. The shock wave travels about four times as fast through rock as through overburden. As shown in Fig. 3.6, a series of primary waves is set up by the shot. Each wave travels through the overburden, and as it encounters new material (rock), part of it is refracted to the surface. By plotting (in milliseconds) the difference between arrival of the primary and refracted waves at the seismometer and the distance (in feet) between the shock and the seismometer, it is possible to determine the depth to bedrock surface.

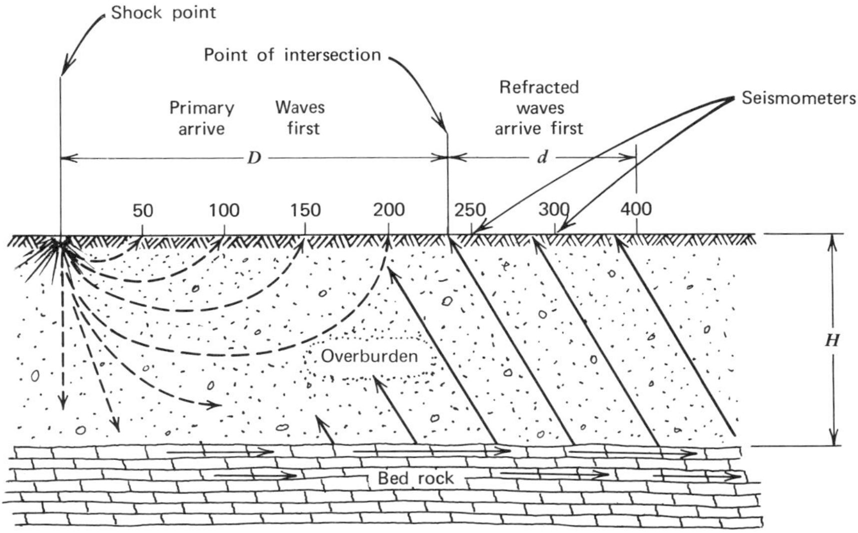

Fig. 3.6 Schematic seismic setup.

The calculation can be made with this formula:

$$H = -\frac{D}{2}\frac{V_2 - V_1}{V_2 + V_1}$$

where H = distance of point of intersection of primary and refracted waves from point of impact in feet;
D = distance to rock in feet;
V_2 = wave velocity in bedrock in feet per second = d/t;
V_1 = wave velocity in overburden in feet per second or D/T.

A review of Fig. 3.7 indicates that each material has its own characteristics, which can be identified or interpreted by a trained operator. Generally, the denser the material the higher the seismic velocity.

Seismographic equipment and methods are used to monitor the effects of blasting on certain construction sites. Adjacent buildings can be protected by regulating the quantity and timing of explosives so that safe seismic readings are not exceeded in these structures.

2. The electrical resistivity method employs electrical equipment to predict the subsurface profile. It is based on the measurement and recording of changes in electrical conductance of various soils. This is accomplished by driving electrodes into the ground and imposing an electrical field. The drop in potential is then measured at points within this field and gives an indication of soil resistivity.

Figure 3.8 shows a schematic setup for resistivity testing. A direct current flow is imposed by a battery between the two outer electrodes, A1 and A2. This current is recorded in milliamperes by a milliammeter. Intermediate electrodes B1 and B2 are equally spaced between A1 and A2, and the potential drop between B1 and B2 is measured in volts. With current, voltage, and electrode spacing known, resistivity in ohm-feet is calculated from the formula.

$$R = 2\pi D\frac{V}{I}$$

where R = distance between electrodes in feet;
D = resistivity in ohm-feet;
V = difference in potential between electrodes B1 and B2 in volts;
I = current flowing between electrodes A1 and A2 in amperes.

Each substratum has its own resistivity, depending on compaction and moisture content. A plot of the resistivity for a continually increasing spacing supplies the data needed to predict accurately the depth and extent of the overburden and the profile of the rock substrata.

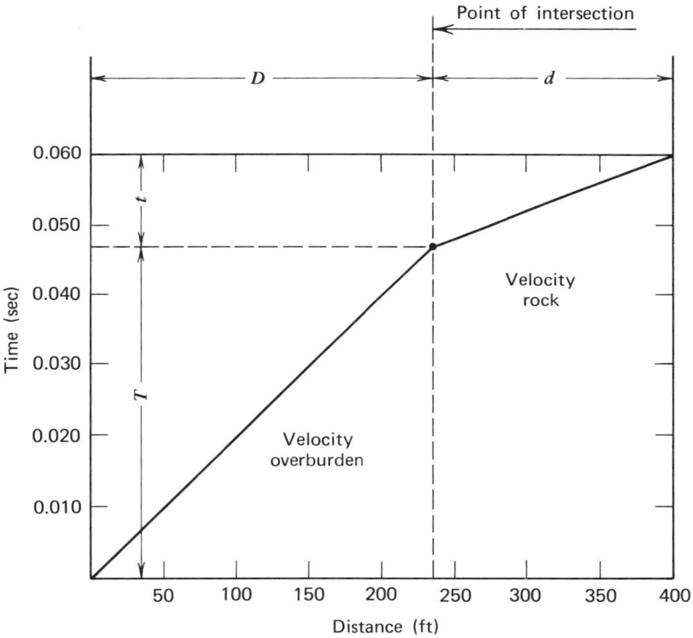

Fig. 3.7 Seismic velocity plot.

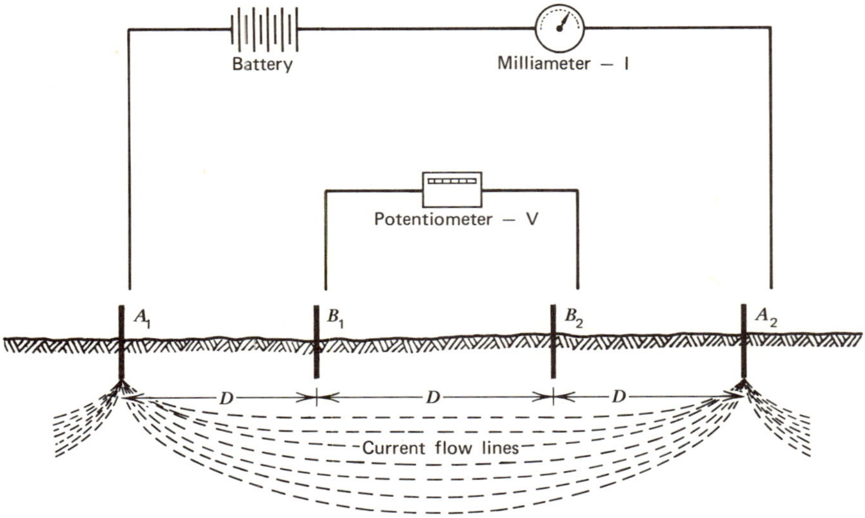

Fig. 3.8 Schematic resistivity method.

The value of both these methods is in their simple equipment and their application to wide areas and large surveys. In both cases, the usefulness of the information recovered is greatly increased by occasional borings and acute observation.

3.4 *Bidding Information*

Whatever the method of investigation, the information must be presented to the respective bidders in an intelligible manner. Soil and rock borings should be shown independently and in profiles to present the information needed for a complete analysis of the substrata. Figure 3.9 shows a finished boring log, which would form a portion of the bid documents. Figure 3.10 shows a soil profile that would be particularly useful in connection with highway work.

There is no standard form for reporting results of soil surveys or foundation borings. Generally, the information required is the same, but the method of reporting it often varies with the individual preparing the report. The quality of all rock samples or cores starts in the field, and depends on two factors: good equipment and competent personnel.

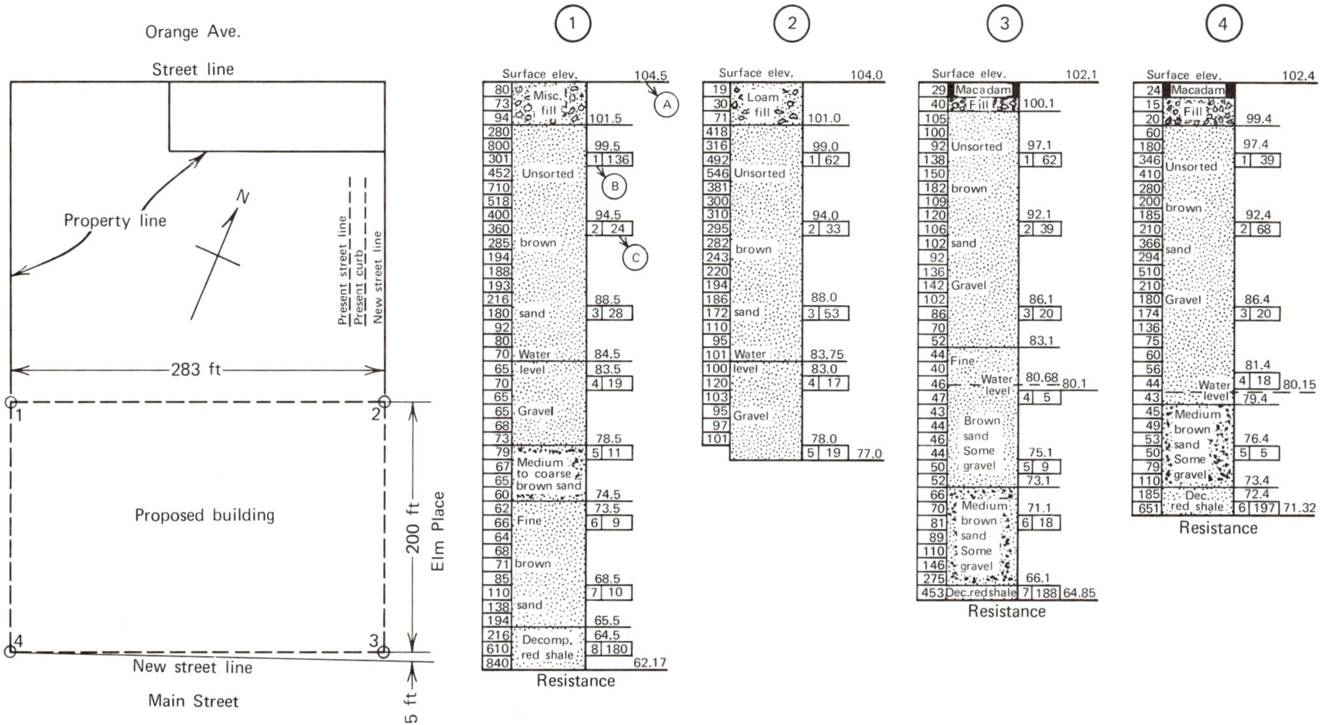

Fig. 3.9 Completed permanent bore hole log. A, Elevation of bottom of sampling device at start of penetration. B, Number of sample or drilling run. C, Number of blows of 300-lb hammer falling 18 in. taken to drive 2½ in. spoon sampler 1 ft. At left of each section is number of blows (same hammer and fall) taken to drive 3-in. casing each successive foot. Record of borings made by Construction Drilling, Inc., New York, for Universal Corporation, Store Building, Plainfield, N.Y.

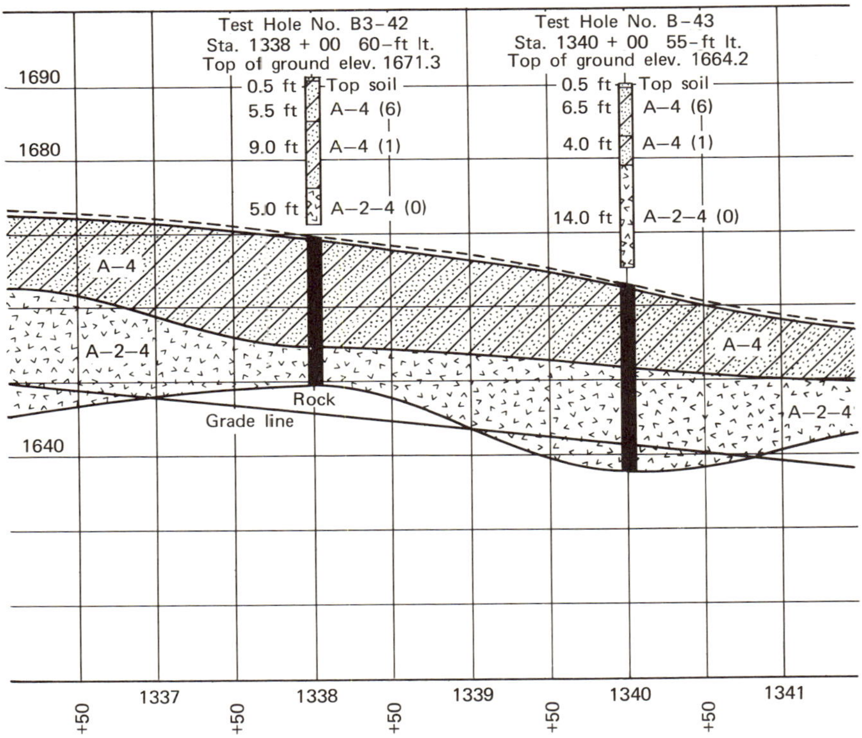

Fig. 3.10 Soil profile, as determined by plotting the borings.

3.5 *Air Compressors*

Air compressors are said to be the basic block on which a rock foundation is built. A faulty compressor certainly idles most of the contractor's work force on a given project.

Air compressors are used to power rotary drills, pneumatic hammers, sheeting hammers, and sump pumps. Every work gang depends on the compressor to provide power for completing the assigned work. (See Fig. 3.11)

When starting a rock foundation or trench excavation, a compressor or group of compressors should be set up to handle the peak demands. The capacity should exceed the amount required to power the air tools by a minimum of 30 percent. With a larger volume than needed, there is less outlet pressure fluctuations and, with an air receiver in the line, the compressor cuts in and out less frequently, thus increasing the life of the machine. An excess air capacity in the receiver or tank on the machine cools by virtue of its staying in the tank or receiver. Cool air holds less moisture and improves the performance of the pneumatic tools.

Air compressors are expensive, renting for about $100/day without an operator. Prospective bidders should be sure of the availability of compressors.

Compressor breakdowns are rare because of the dependability of the machine, but normal service is required to maintain this dependability. Good maintenance personnel must be available for repairs under emergency conditions.

A typical air compressor is shown in Fig. 3.12. Heavy-duty air compressors providing up to 1200 cfm are standard, and manufacturers provide literature on request.

Air compressors that are to be positioned along city streets are subject to city ordinances regarding noise and air pollution. The exhaust can be baffled through mufflers and purified through filters. However, these methods are notorious for being inadequate for the large compressors. The only successful method is to enclose completely the compressor with a temporary building constructed of a sound-absorbing material or use the latest Whisper models which feature their suitability to urban environments.

Air compressors can have a variety of secondary functions, such as providing the power required by air pumps (Fig. 3.13) and miscellaneous tools.

Fig. 3.11 Compressed air is used to activate these chipping tools. Close quarters result in the extensive use of hand tools in rock excavation. Cost/yd^3 may be of the order of ten times the bulk-removal rate.

Fig. 3.12 Air Compressors may be series connected to assure trouble-free performance and adequate supply.

Fig. 3.13 Pumps operated by air compressors are handy for dewatering pier and footing shafts.

3.6 Air Tools

Air tools vary in size and requirements, from the jackhammer with a requirement of 50 cfm to the Ho-Ram or jumbo drill, which may require over 1000 cfm. Tunnel men may be familiar with rotary drills that require many times this capacity to operate a multitude of drills simultaneously.

Standard types of drills and air tools are illustrated in Figs. 3.14 and 3.15. Additional information can be obtained from the manufacturer.

Cat or wagon drills (as manufactured by both Ingersoll-Rand and Gardner-Denver) (Fig. 3.16) have proven to be rugged performers that withstand the common abuses inflicted by construction personnel. These drills have ample power and speed to handle any hole size from 2½ to 4 inches. They are designed to drill angle holes, blast holes, and line holes in areas previously thought to be inaccessible to standard equipment. Heavy-duty traction motors are capable of propelling the drill and a compressor up steel slopes. (See Figs., pages 74–88).

The cat drill is truly a workhorse for all types of rock foundations. The drill rod is the most abused part of the equipment because it delivers the drill's power to the rock substrata. To be effective, a rod must be simple to handle and the bit must be tough enough to grind up the rock. Figure 3.17 shows the various components in a typical drill rod.

Manufacturers continue to improve on the capability of their drills. Figure 3.18 features a rotary drill capable of drilling 8-in diameter blast holes under a variety of conditions.

Fig. 3.19 & 3.20 (W51, W52) indicate excavation operations at typical sites.

3.7 Tunneling Machines

Giant moles are capable of drilling 21-ft diameter holes through rock formations. These machines are guided by laser beams and have a positive anchorage system that enables them to propel themselves through difficult strata. In one system (Fig. 3.21) a pilot bore is used as an anchorage, and the rotating face is literally pulled into the rock. Another system (Fig. 3.22) utilizes a side anchorage that pushes the tunneling machine into the rock formation.

GARDNER-DENVER HAND HELD

SINKER DRILLS

Gardner-Denver drills are designed with the customer in mind. Maintenance costs are reduced by the generous use of replaceable bushings, sleeves and reversible parts. Several models incorporate built-in tubular air screens which are readily cleaned by a flick of the control handle. Here too—the use of a special gland design in the backhead permits changing the machine from dry to wet to automatic-wet constructions merely by changing the gland and tube—without taking the drill apart. This same design eliminates the difficult job of tube removal, since the tube is withdrawn with the gland. Reliable, inexpensive "O" ring is used in many places for sealing air and water.

	S17H	S33	S48	S55	S58
Piston Diameter	1¾″	2⅜″	2½″	2⅝″	2⅝″
Overall Length (with Collared Chuck)	18½″	21″	22″	24″	23″
Total Weight (Dry Construction)	26¼ lbs.	33¾ lbs.	51¼ lbs.	59 lbs.	62 lbs.
Air Hose Size	½″	¾″	¾″	¾″ or 1″	¾″ or 1″
Water Hose Size (Wet Construction)	⅜″	½″	½″	½″	½″
Chuck Sizes	¾″ and ⅞″	⅞″	⅞″ and 1″	1″	1″ and 1¼″
Shipping Weight (Dry, Collared Construction)	31 lbs.	41 lbs.	60 lbs.	64 lbs.	73 lbs.
Shipping Dimensions (Carton)	5″x7″x24″ 1 cu. ft.	5″x7″x24″ 1 cu. ft.	5″x7″x27″ 1 cu. ft.	5″x7″x27″ 1 cu. ft.	7″x7″x29″ 1 cu. ft.

Fig. 3.15 Costs for rock excavation may be reduced considerably by using modern tools and attachments.

Fig. 3.16 Howard Johnson's Motor Lodge on 52nd Street and 8th Avenue in Manhattan, 1961. Wagon drills are used to line drill adjacent to existing structures. Dust machines are required to filter the rock dust by means of a direct hookup to the wagon drill. Rock workers may become fatally ill if they breathe a dusted atmosphere for extended periods.

GARDNER-DENVER

SECTION 8-8H
1ST EDITION

TUBE-ROD

New SERIES OF TUBULAR DRILL ROD
FOR DRILLING OF DEEPER AND LARGER DIAMETER HOLES

Tubular rod diameter sizes 2½"-3"-4"

The new series of tubular drill rod equipment is specifically designed for larger diameter and deeper hole drilling. While the thread form is similar in design to the Hi-Leed® drill rod, it has a larger thread slope and increased lead for ease in uncoupling.

You have these advantages when drilling with the new Tubular Drill Rod . . .

- **BROKEN GROUND**

Drilling broken ground where cutting removal air is lost and hole deviation encountered with conventional sectional steel is excessive to the point the drill string becomes locked. In such a situation the tubular steel affords better hole cleaning because of reduced annulus between steel and hole wall. It also is less flexible thereby maintaining good alignment.

- **HEAVY FORMATIONS**

Drilling heavy formations where cutting removal is difficult because of insufficient cleaning air velocity. The reduced annulus and smooth flush jointed connections favor higher cleaning air velocities to carry out the cuttings.

- **LIMITED AIR**

Drilling situations where air availability is limited. Reduced blowing air requirements may represent the margin needed for successful drilling and cutting removal.

Here are the distinct advantages of Tubular Steel over Sectional Rods . . .

1 STRONGER TUBE:

A. Larger threads which give:
1. more thread contact;
2. easier uncoupling;
3. less coupling heat.

B. Larger component parts which include:
1. shank;
2. coupling;
3. bit adapter.

2 STRAIGHTER HOLES:

A. More rigidity of tubular rod results in:
1. less drift of bit;
2. longer "string" of drill steel.

B. Larger diameter tube which resists:
1. bending;
2. twisting.

3 BETTER HOLE CLEANING:

A. Smaller annulus area allows:
1. larger dia. holes;
2. deeper holes;
3. better cutting removal;
4. more air velocity.

B. Smooth outside walls resist:
1. binding of bit;
2. interference with flow of cuttings.

GARDNER-DENVER Since 1859

GARDNER-DENVER COMPANY
QUINCY, ILLINOIS

Fig. 3.17 Drill steels are manufactured in varying lengths and sizes, in order to increase their flexibility.

Fig. 3.18 New T-4 Drillmaster. World's first dual-purpose rotary drill.

Fig. 3.19 74th Street in Manhattan. Incinerator, 1954. Shovels, trucks, and wire mats are all part of an efficient operation. Equipment such as the above lends itself to jobs where a daily production rate of over 300 yd^3 is anticipated.

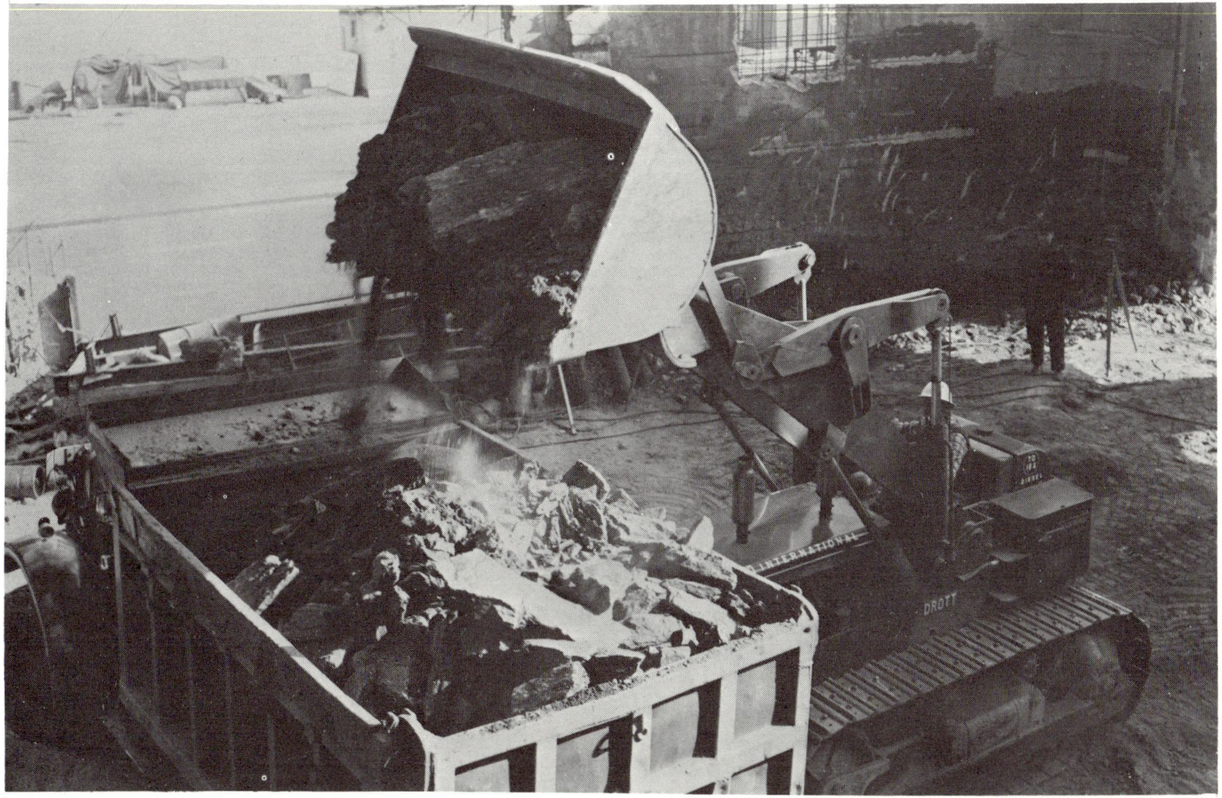

Fig. 3.20 Incinerator on 74th Street in Manhattan, 1954. Front end loaders are less expensive than shovels and are operated by one man. They perform an efficient job if the material size, after blasting, is less than ½ yd^3 in volume.

Fig. 3.21 The Alkirk Hardrock Tunneler delivers more than 1 million lb of thrust. The above was built for Perine-Morrison-Knudsen, joint venture contractors on the New York City Board of Water Supply's Richmond Tunnel and was designed on the basis of a 12-ft-diameter bore.

Fig. 3.22*a* Tunneling machines have revolutionized the industry.

Fig. 3.22*a, b* The Jarva Mark 21 provides advantages such as 21-ft-diameter boring capability; laser guidance system; hydraulic starting motors; automatic lubrication of critical parts, including cutters; ability to erect supports or roof bolts near the face; enclosed operator's cab; free access to the face; negative pressure dust control.

Air Compressors may be used in a variety of ways and should be situated in convenient locations.

Air tools must be kept in good working order by trained mechanics.

Flexible bull-hoses are used as connectors to power large drills.

Line holes may be inclined to secure stable cliffs.

Paving breakers are cushioned to provide maximum breaking power.

Wagon drills are rugged and may be used under the most severe conditions.

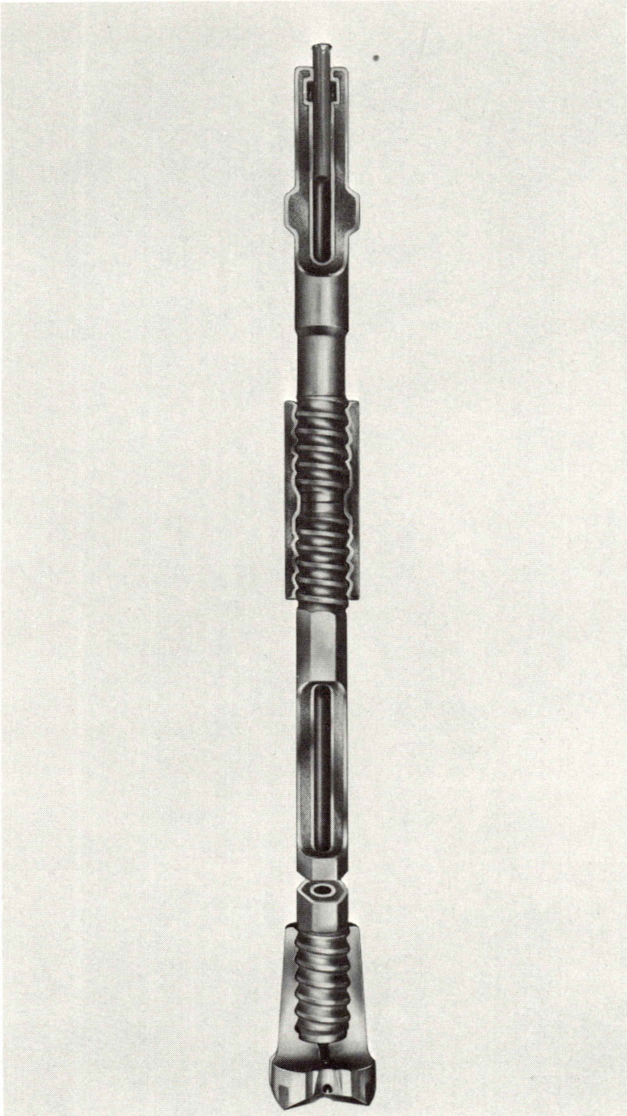

Drill steels and bits are designed to deliver top performance.

Wagon drills have hydraulic arms that enable them to position the drills in otherwise inaccessible areas.

Natural rock formations are rarely level or uniform.

Jumbo Drills are capable of high performance.

Rock drills must be properly positioned for maximum effectiveness.

Stationary drills with fixed leads may be used for coring rock.

Core samples, acquired at various depths, enable designers to estimate probable rock strata's.

Sampling devices mounted on trucks provide's long range mobility.

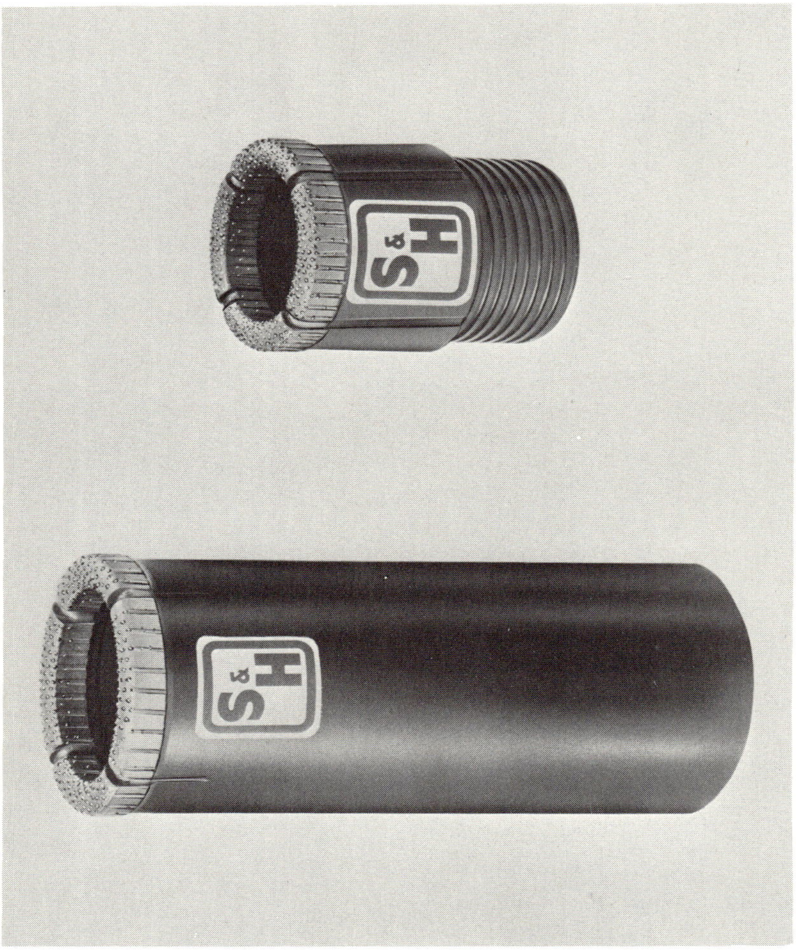

Core bits are available in varying sizes and hardness.

Muck buckets located on the periphery of the cutterheads constantly clear the grindings from the head. The uniform chips are picked up at the invert and carried in the buckets to the top of the cutterhead, where they spill into a conveyor that carries them to the rear for removal from the tunnel.

4

Blasting

4.1 Breakage

Blasting as a method of fracturing a rock formation has as its chief advantage a high concentration of power over a relatively short span. This burst of energy provides a concentrated pressure that shatters the area adjacent to the drill hole and exposes the space beyond to vast tangential strains and stresses.

An exposed face or bench provides the expansion path required to shatter the fragments. As the burden in front of the drill hole is removed, tangential stresses are released in adjacent areas causing spalling and cracking. The early movement of the rock face is most important for an efficient blast, which achieves maximum cracking per drill hole with the minimum quantity of explosive.

Figure 4.1 demonstrates the probable fracture planes at the bottom of the bench under varying conditions. Obviously neither example would be satisfactory over extensive areas. The free bottom saves on drilling and loading costs, and it may be used advantageously whenever the excavation is to be completed in successive cuts or benches. The fixed bottom would be used to remove the final bench. This method saves on come-back costs, which tend to build up if the toe of the slope is not blasted properly. The rock breakage in blasting one or more rows of drill holes depends largely on the relation between the burden and the hole spacing, and also on the time of ignition between holes. When ignition occurs simultaneously, gas pressure in several holes cooperates and pushes the rock forward as the free face yields. If delays are used, each individual hole must tear off a part of the burden. This is one of the major reasons for the improved breakage obtained in short-delay blasting. (See Fig. 4.2 to 4.5.)

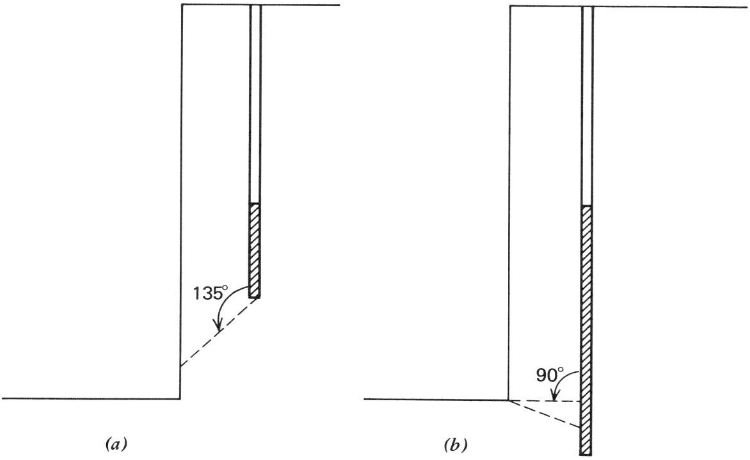

Fig. 4.1 Natural angles of fracture. (*a*) Indicates an angle of 135°, whereas (*b*) demonstrates the importance of the free-face principle.

Special equipment commonly used to fragment rock beds by the use of explosives must be purchased by the contractor or his agent to assure a safe and satisfactory job completion.

Commercial high explosives are commonly known as dynamites. There are several types of dynamites in regular use, and each type is subdivided into a series of grades. Some of the principal variables of these grades are strength, density, sensitiveness, velocity, water resistance, freezing resistance, and fumes as products of detonation.

The major suppliers of explosives have available a dependable service that provides reliable recommendations for the type and grade of dynamite best suited for the individual job. These representatives should be fully informed of the requirements of the job and continually consulted during the progress of the job. The term "strength" refers to the energy content of an explosive, which determines the force and power it can develop and the work it is capable of doing.

Nitroglycerin, or straight dynamites, are rated in strength according to the percentage by weight of their nitroglycerin content. A 20-percent straight dynamite actually contains 20 percent of nitroglycerin by weight. A 60-percent straight dynamite produces one and a half times the energy output of a 20-percent-strength dynamite. When the percentage of nitroglycerin is increased to 60 percent, the energy obtained from the nitroglycerin alone is trebled, but that contributed by the remaining ingredients is reduced.

Fig. 4.2 Incinerator on 73rd Street in Manhattan, 1954. Blasting is a highly specialized profession and can be quite safe if standard precautions are taken. The blaster, or dynamite specialist, should be present while the hole is being drilled and certainly during loading.

Fig. 4.3 Princeton Club on 43rd Street and 5th Avenue in Manhattan, 1961. Sheeting tied back into rock formations by grouted rods simplifies the blaster's task because he is not required to reduce his charges in the berm area. Interior braces complicate everyone's job and should be avoided when possible.

Fig. 4.4 777 Third Avenue in Manhattan, 1962. Blasting to remove rock berms must be carried out with care. Workmen should wear safety hats at all times.

Fig. 4.5 777 Third Avenue in Manhattan, 1962. Overbreak frequently cannot be prevented even though line drilling has been properly employed. Experienced blasters and superintendents continually scale loose and insecure blocks. Some cases may require the use of rock bolts and chain-link fence for safety.

An alternative to nitroglycerin is the use of insensitive chemical mixtures that explode when initiated with a high-explosive primer. These mixtures are known as "blasting agents," and have the advantage of added safety in handling and lower costs.

Igniters or safety fuses convey fire at a continuous and uniform rate for a direct firing of an explosive charge, as in the use of black blasting powder, or for indirect firing, as in the ignition of a blasting cap to detonate dynamite. A safety fuse is readily susceptible to damage, and is often the cause of faulty performance.

Blasting caps are small tubes or shells that contain a highly sensitive explosive. They are detonated by the spit or sparks from a safety fuse. Blasting caps are dangerous and should not be tampered with in any way by inexperienced personnel. Electric blasting caps are fired by means of an electric current. They, too, are subject to detonation from impact or rough treatment.

Delay electric blasting caps have a delay element inserted between the electrical firing element and the detonating charges. Delays are used to fire charges of dynamite in rotation, and their advantages over the use of caps and fuse for this purpose are greater safety, great convenience, and more accurate timing. Rotation firing has the general advantage that complete rounds can be shot in one operation.

The detonating fuse is essentially a small round core of explosive covered with a protective sheath. It is attached to the first cartridge loaded into the hole and acts as a priming agent throughout the entire length of the column of explosives.

Blasting machines supply current for firing blasts by electricity. There are two general types: one type uses dry-cell batteries to charge a bank of condensers that supply a direct current of short duration to electrical firing devices; the second type is a generator that delivers a direct current. Both are rugged and withstand hard service for long periods.

The generator, or push-down, type has a plunger capable of delivering enough energy, when depressed, to fire 50 or more blasting caps.

Testing instruments, or galvanometers, are used by the blaster who is testing individual electric blasting caps to determine whether or not a blasting circuit is closed and in proper condition for firing; locating broken wires, faulty connections, ground and short circuits; and finding the approximate resistance of a circuit.

A combination voltmeter and ohmmeter is also used for checking stray currents and for reading line voltages.

Blasting mats are used to cover loaded holes just before firing to catch or hold material flying from the blast. These heavy mats may be constructed of sisal rope or of steel-wire rope or chains. If they are to cover heavy charges, railroad ties or logs should be put down first and the mats on top of them. The woven steel mats are most effective for holding down rock fragments from a blast. Care must be exercised in placing the mats because short circuits are likely if bare connections touch the wire rope.

The storage of explosive on construction sites is usually governed by law, which requires the use of separate sand-bagged magazines for the storage of blasting caps and dynamite. Magazines should be secure from pilferage or tampering. In metropolitan areas this may mean the employment of specially trained guards to safeguard the explosives around the clock. It may be practical to have explosives delivered and used on a daily basis; that is, the blaster receives or picks up daily only the amounts necessary for each day's activities. These explosives are secured during the day in magazines under his care. The magazines, if empty at night, are then left open as a sign of safety.

4.2 *Calculation of the Charge*

Determination of the drill-hole pattern, size of the charge, and sequence of breakage are the main problems of practical rock blasting. In standard cases with a repeating pattern, the quantity of the charge can easily be increased if previous shots have proven insufficient. "Calculation of the charge" basically means calculation of the *minimum* charge required to loosen or fracture the rock. This minimum charge must be refined to take into account the effect on surrounding rock and/or neighboring buildings.

Figure 4.6 illustrates typical charges for normal homogeneous-rock conditions. These charges should be modified as work proceeds and fragmentation is observed.

Rock excavations must be planned with the inherent characteristics of the job in mind. Bench height must be based on the overall cut and the ability of the equipment to drill at a reasonable cost. In populated areas bench heights are usually limited to 12-ft cuts because larger benches are difficult to cover with wire blasting mats, and the added explosives tend to throw small stones great distances.

The charge to be used may be calculated as roughly equal to 0.5 lb of dynamite per cubic yard of rock to be removed. A few trial blasts determine the extent to which this quantity should be modified. Larger concentrations of charge decrease the size of fragments and increase the throw of particles.

The covering of a blast is extremely important, for adequate mats must be provided to cushion the energy that has been transmitted to the individual rock particles. Scattering of stones or rock fragments after a blast is difficult to control and often can break windows in adjacent buildings. Piling mats on the material to be blasted tends to screen out the smaller particles, which can fly as far as 100 yd from the explosion.

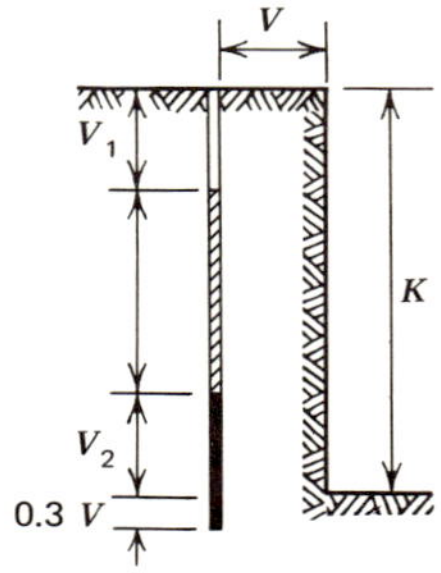

(*a*) Ordinary benches ($K > 1.8\ V$)

Typical Examples

K = Bench height = 20 ft
d = Diameter of drill bit = 2 in.
s = Weight strength of explosive dynamite with 35% nitroglycerin = 1
V = Burden = 7.5 ft = $V_1 = V_2$
E = Spacing = 1.25 V = 9.3 ft on center
lb = Concentration of charge = 1.5 lb/ft at the bottom
lp = Concentration of charge = 0.6 lb ft in the column

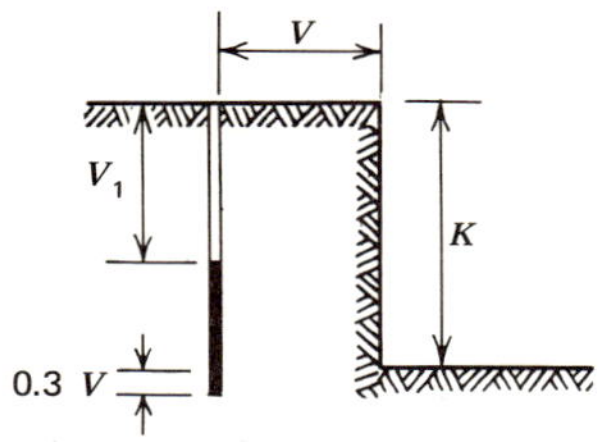

(*b*) Low benches ($K < 1.8\ V$)

K = 12.5-ft bench
d = 2-in. drill bit
V = 7.5-ft burden = V_1
E = 1.25-V = 9.3-ft spacing
Q_t = 12 lb

$$l = \frac{12}{5 + 0.3\,V} = \frac{12}{7.25} = 1.65 \text{ lb/ft}$$

Note. Low bench requires no column charge because of the shallow depth. In all cases the upper portion of each drill hole should be stemmed with dry sand.

Fig. 4.6 Typical charges for normal and low benches.

4.3 Loading Drill Holes

Filling the drill hole with a dense explosive gives a higher yield per pound of explosive used. A tamped or compacted drill hole must first pressurize all voids with gas before any energy is actually transmitted to the rock formation. Despite the inherent disadvantage of manually tamping each cartridge, this practice continues in most major cities because of antiquated building codes.

The standard procedure for loading a drilling pattern has been to use a wooden rod for tamping the explosive. When compressed the cartridge expands, filling the drill hole. The cartridge often is perforated or slit to facilitate this forced expansion. This is a primitive method, and experience shows that the average degree of packing is comparatively low. (See Fig. 4.7.)

Compressed air has recently been adopted for blowing cartridges into drill holes. The cartridges travel so fast that they are crushed when striking the bottom.

Pneumatic cartridge loaders give a high degree of packing and enable contractors to increase the spacing of drill holes, which decreases the amount of dynamite required to move a cubic yard of material.

4.4 Methods—A Case in Point

Blasting methods in metropolitan areas are limited by restrictions such as were imposed on the construction of the East 74th Street Power Plant of the New York City Transit System. Specifications stated that "No blasting will be permitted except where express permission has been granted by the engineers after receiving a detailed request in writing from the Contractor." Preliminary meetings were held with the plant operating personnel, engineers for the designers, and the owners to discuss warning signals, inspection, and control of the shooting. (See Fig. 4.8.)

As a result of the planning sessions, permission to blast was granted, subject to strict compliance with 19 conditions. These included a warning system for plant personnel; restrictions on blasting during the morning and afternoon peak loads of subway rush hours; use of a vibrograph to record the effect of each blast; limitation to ½-lb sticks of 40 percent DuPont Special Gelatin with millisecond-delay blasting caps; checking of the site for stray current before delivery of caps; use of three 10 by 12-ft steel-cable blasting mats as minimum cover for each blast; and specifically stating the explosives loading schedule.

The initial allowable load was stated as a minimum of three shooting holes loaded with a maximum of one and one-half sticks, or ¾ lb, of dynamite. Increase in charge for a specific load area was restricted to a maximum increment of one-half a stick subject to the approval of the engineers, based on vibrograph indications under similar conditions.

The entire problem was carefully considered by both owner and contractor. In addition, the contractor brought in an independent seismologist to help set up and calibrate the control program using seismographs.

As shown in Fig. 4.9, the working area (110×75 ft) is wholly within the power plant, and one end of the tunnel excavation is 10 ft from an operating turbogenerator. Work access to the job was through a door at street level at elevation +16 opening onto 74th Street. The door was 19-ft wide with a 16-ft clear height.

Fig. 4.7 Loading drill holes in highly congested areas, such as central business districts, requires special care and expertise. Explosives are loaded and tamped using a wooden safety rod.

Fig. 4.8 Extensions of discharge tunnel on left and intake tunnel on right carry cooling water for additional turbine generator. Note line drilling and broaching for sides of tunnel.

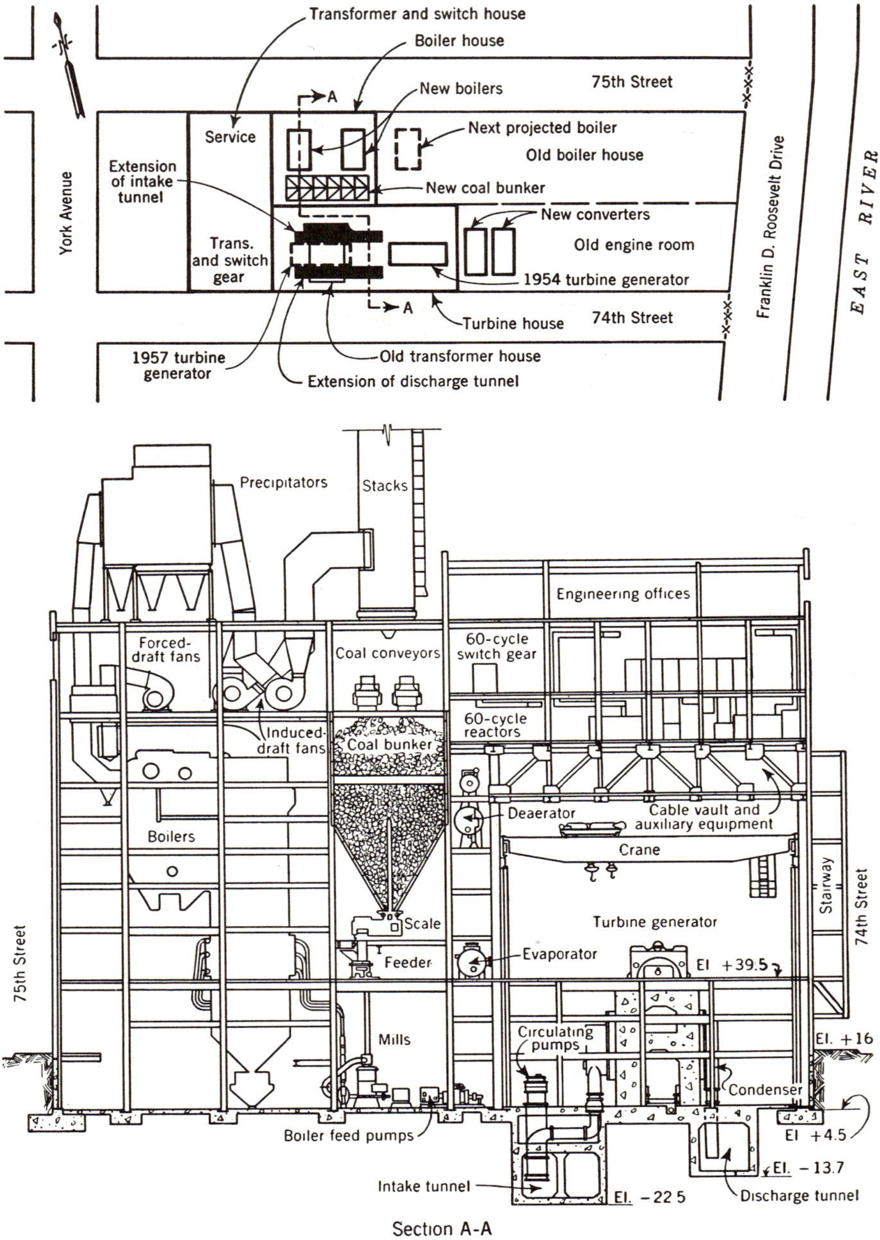

Fig. 4.9 Installation of 1957 turbine generator. Note its proximity to existing turbine generator and other important parts of the operating power plant. This drawing is revised from that printed in November 1954, *Civil Engineering*, p. 57.

At the start of excavation, the 70 × 80-ft foundation of the old transformer house was still in existence below elevation +16. The remaining peripheral area was at elevation +4.5 (the old basement floor).

The mass of concrete, earth, and rock under the old transformer house was excavated to the +4.5 level using a 2-yd^3 shovel that was ramped down to the lower level. Trucks were backed down the earth-rock ramp and loaded on the lower level. As rock was encountered, the careful program of blasting control was instituted.

A Shrader tridimensional vibrograph operated throughout the rock excavation work, measuring and recording each blast. As a direct-reading instrument, the vibrograph was positioned for each blast at various sensitive locations, including the adjacent turbogenerator, unit substations, the mechanical control room, and adjacent boiler-feed areas. Individual complaints of excessive vibration, shock, or noise were reviewed and investigated, often by positioning the vibrograph on the desk of the person making the complaint. Through a cooperative effort by plant operating personnel, engineers, and the contractor, loading of the holes was controlled for various locations of the working area, and continually varied as the depth of rock excavation increased.

The nearest vibrograph checkpoint, the adjacent turbogenerator, showed less vibration effects from the blasting than was found in the plane of rotation of the generator's moving components. The area most sensitive to blast effect proved to be the meters in the mechanical control room, the needles of which acted as low-order seismographs.

The blast effect studied in the control program was the dynamic displacement as measured by the vibrograph, not the more common measure of noise level or shock as perceived by a person seated nearby.

The shovel cut was continued to elevation 0.0, the bottom of the turbine mat, in the north half and southwest quarter of the working area. This allowed the line drilling and broaching operation for the intake tunnel to start, thus maintaining job continuity. At this stage, the shovel loaded trucks standing on the ramp at the southeast corner; finally, it loaded them standing just inside the door as the ramp was blasted out. To remove the last piece of the rock mass in front of the door at the southeast corner, the shovel, which could not reach truck height, loaded the bucket of a rubber-tired front-end loader positioned at the doorway on the street. The loader, in turn, loaded the trucks on the street.

Throughout the operation of excavating to elevation 0.0, the line drilling operation was carried out in two shifts following closely behind the excavation. At the completion of the shovel-to-loader-to-truck phase of the work, the Lorain shovel was converted to a crane with a 60-ft boom. In addition, a platform was constructed at street level with supporting bents straddling the discharge-tunnel excavation.

With the crane positioned between the tunnels, blasting began in both tunnel cuts, working from west to east. The crane placed protective mats over the loaded shooting holes and also dug and loaded rock muck into trucks on the platform at the doorway. The crane functioned in this manner until a shallow cut had been made in both tunnels.

To complete the tunnel excavation the next and final equipment change was to rig the crane with a three-part cable to handle efficiently heavier lifts and to bring in a 2½-yd^3 Traxcavator front-end loader to muck the deeper rock cuts, placing the material in 3-yd^3 bottom-dump "battleship" buckets, which were lifted out of the cuts and loaded into trucks by the crane. Both tunnels were worked simultaneously, the front-end loader being lifted back and forth by the crane.

A total of 3200 yd^3 of rock was excavated, using an average of less than 1 lb of dynamite per cubic yard.

Although most jobs do not require the intricate procedures outlined above, it is wise to realize that all rock excavations require prior planning to avoid accidents and excessive costs.

4.5 Ground Vibrations

Rock excavations in central areas require close supervision and monitoring of the vibrations inaugurated by the blast. When blasting work has to be carried out in the neighborhood of existing structures, ground vibration is often the factor that finally decides how the blasting is to be planned (Fig. 4.10). Experience has shown that the great majority of vibration complaints and law suits arise from physical and mental irritation that, in turn, provokes imaginary damage to a dwelling or its contents (Fig. 4.11). Noise and air concussion and actual ground movement are all less noticeable when people are going about their daily pursuits during the day rather than evenings or weekends (Fig. 4.12).

If ground vibration problems are likely to arise, an estimate of vibration effects should be considered. The risk of damage determines the charge size and the distance between the charge and structures. In hard rock, a single 20-lb blast can cause cracking or loosening of plaster at a distance of 150 ft, whereas a 2-lb charge may be perfectly safe at 30 ft. Smaller charges with delays maintain the vibration level at a safe value when close to structures (Fig. 4.13).

The standard procedure in metropolitan regions is to position a vibrograph in areas considered susceptible to blast damage and actually monitor the vibrational effect during the blasting operation. A record of blasting effects compared with size of charge and distance from point of

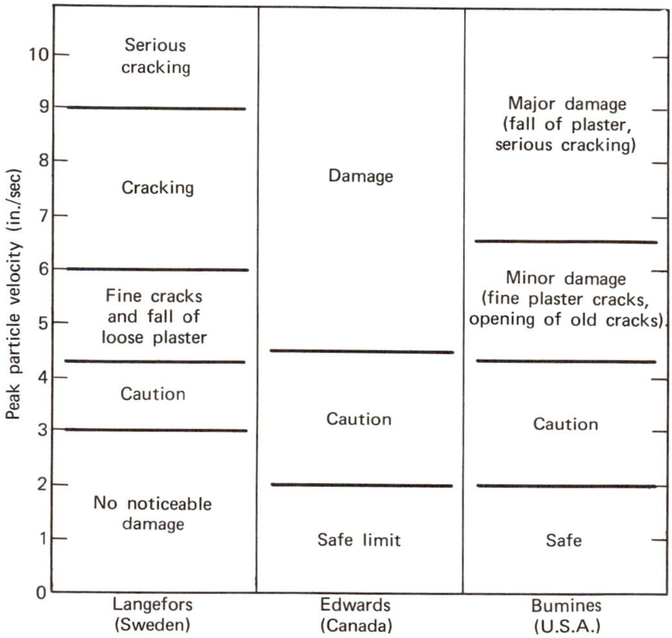

Fig. 4.10 Three researchers, Langefors, Edwards, and Bumines, agree closely on the peak particle velocities at which varying degrees of structural damage will occur. We recommend 2 in./sec as a safe maximum for preventing cracking of plaster in residences.

detonation normally is sufficient to reduce the possibility of damage to a minimum.

4.6 Line Drilling and Presplitting

"Smooth blasting" a rock face has been advanced by recent studies of blasting effects. A series of line holes is drilled along the property line or other dividing line with a spacing of from 4 in. to 2 ft, depending on the burden and the tolerances of fragments projecting into the excavation.

Depth and alignment of drill holes are important, for a careless pattern results in a face that tends to wander from hole to hole. The best faces are achieved when bench depths are limited to under 20 ft (Fig. 4.15).

A concentration of only 10 percent of the amount normally used in a fully loaded round is generally sufficient to cause cracking between line holes. A greater concentration inevitably causes random rosettes of cracks

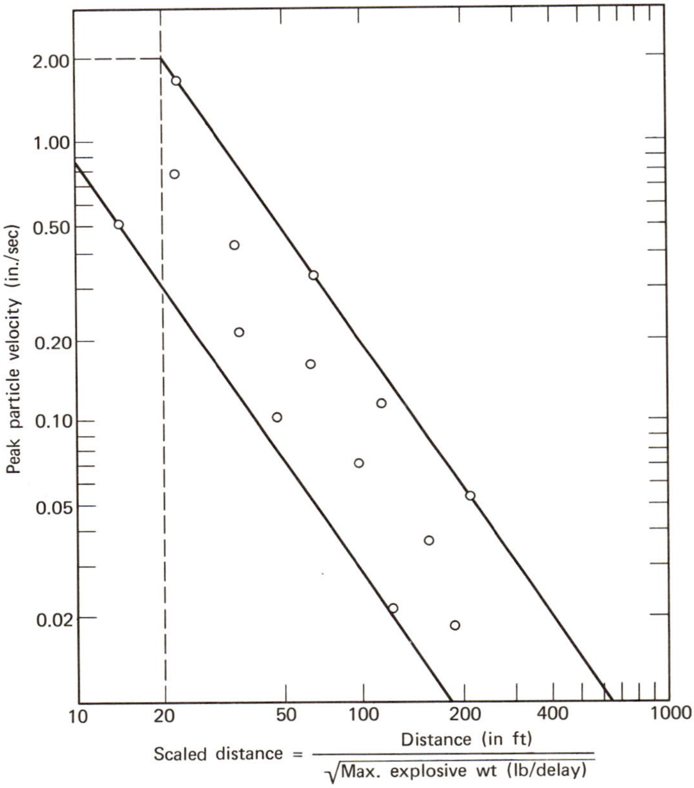

Fig. 4.11 Peak particle velocity (a measure of vibration intensity) is plotted against scaled distance. Note that in this situation, if one wanted to avoid cracking of plaster in homes, he would limit the scaled distance to 20 ft.

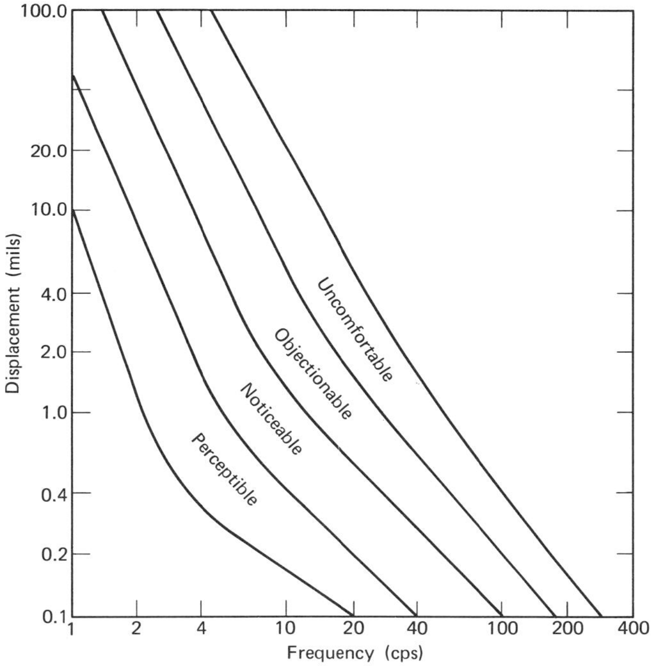

Fig. 4.12 Both frequency and displacement influence subjective human responses to nearby blasting.

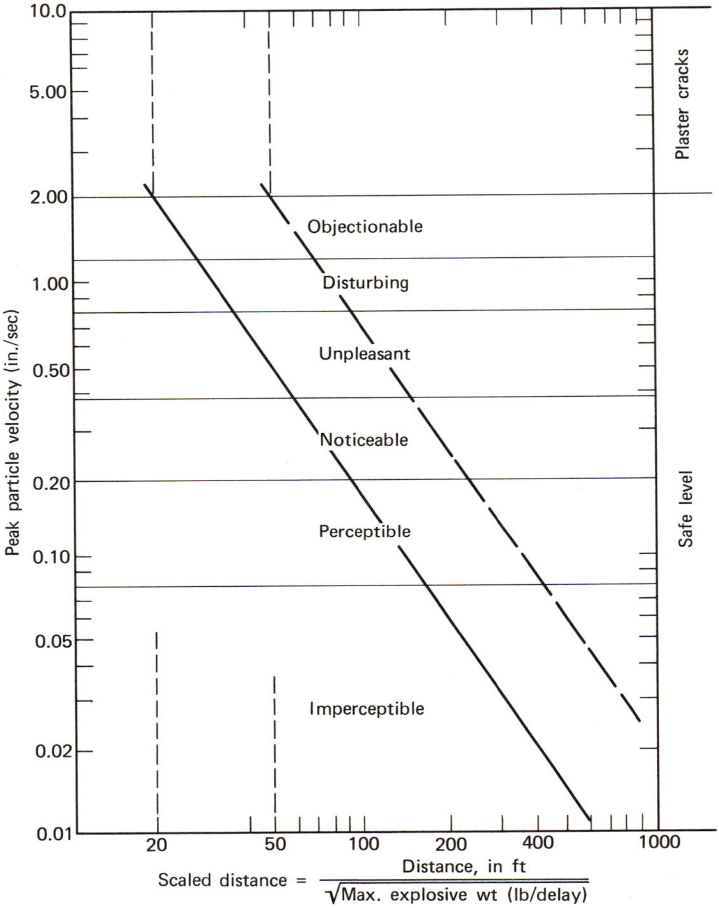

Fig. 4.13 This is Fig. 4.11 with a human evaluation scale added. The solid line is the upper curve of Fig. 4.11. While the structurally safe level is 2 in./sec, note how much lower the peak particle velocity must be to minimize neighbors' complaints. The dashed line should be used to control a blasting operation for which no seismographic tests have been conducted.

Distance (ft)	Scaled Distance 50	35	$D/\sqrt{E}$ 20
50	1	2	6
100	4	8	25
200	16	32	100
500	100	200	650
1000	400	800	2500
2000	1600	3200	10000

Fig. 4.14 Maximum explosive weights (lb) per delay allowed for several scaled distances.

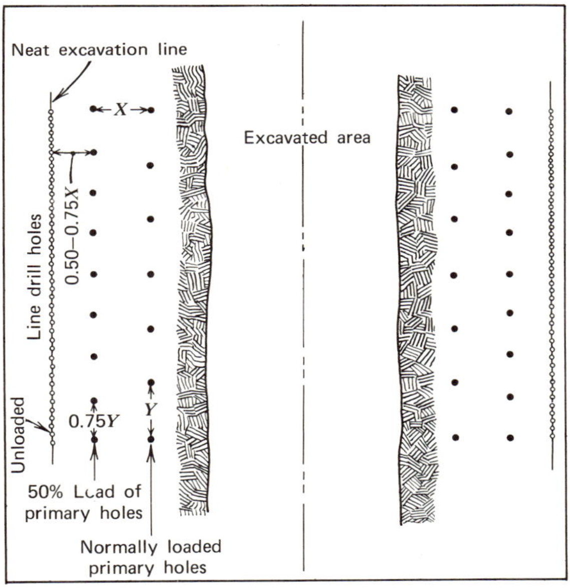

Fig. 4.15 Typical pattern and procedure for line drilling. *Principle.* Line drilling involves a single row of closely spaced, unloaded, small-diameter holes along the neat excavation line. This provides a plane of weakness which the primary blast can break. It also causes some of the shock waves created by the blast to be reflected which reduces shattering and stressing of the finished wall. *Advantages.* Line drilling is applicable in areas where even the light explosive loads associated with other controlled blasting techniques may cause damage beyond the excavation limit. When used with other controlled blasting techniques, line drilling between the loaded holes promotes shearing to improve results. *Limitations.* There are a number of limitations of line drilling which must be recognized: line drilling is rather unpredictable except in the most homogeneous formations; due to the close spacings required, drilling costs are high; because line drilling requires a large number of holes on rather close spacings, drilling becomes tedious and results are often unsatisfactory due to poor hole alignment.

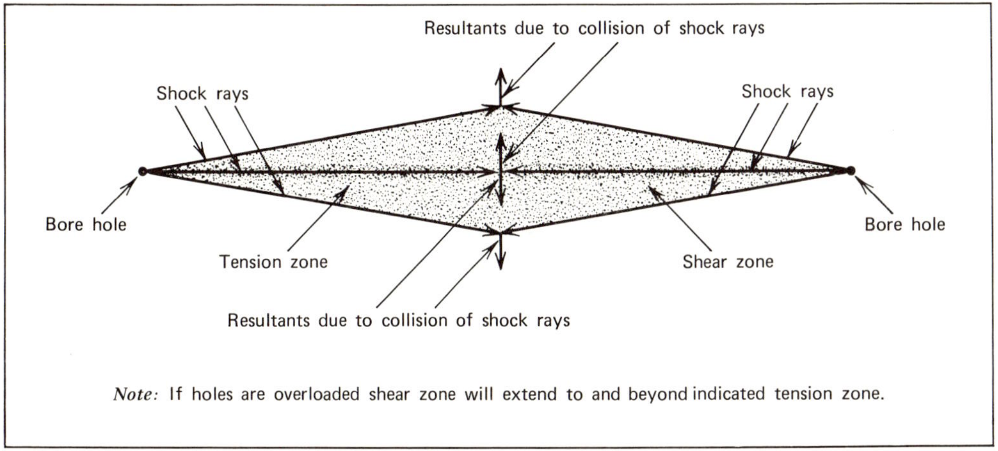

Fig. 4.16 Principle of preshearing. *Principle.* Preshearing, sometimes referred to as presplitting, preslotting, or stress relieving, involves a single row of holes drilled along the neat excavation line. The holes are usually the same diameter (2 to 4 in.) and, in most cases, all are loaded. Preshearing differs from line drilling, cushion blasting, and smooth blasting because the holes are fired before any adjoining main excavation areas are blasted. The theory of preshearing is that when two charges are shot simultaneously in adjoining holes, collision of the shock waves between holes places the web in tension and causes cracking that gives a sheared zone between the holes. With proper spacing and charge, the fractured zone between the holes will be a narrow sheared area to which the subsequent primary blasts can break. This results in a smooth wall with little or no overbreak. The presheared plane reflects some of the shock waves which follow from the primary blasts preventing them from being transmitted into the finished wall, minimizing shattering and overbreak. This reflection of shock waves should also tend to reduce vibration. *Advantages.* Preshearing offers these advantages: increased hole spacing—reduced drilling costs; it is not necessary to return to blasting slopes or walls after primary excavation. *Limitations.* In preshearing it is difficult to determine results until primary excavation is complete up to the finished wall. Since preshearing is done before primary blasts are made, it is not possible to take advantage of the knowledge of local rock conditions which is gained in primary blasts. Also, the hole spacings in cushion and smooth blasting can usually be greater than in preshearing, thus reducing drilling costs.

spreading in all directions, resulting in a weak face that tends to fragment and loosen with time. Normally, this low concentration is implemented by loading alternate holes (Fig. 4.16).

In presplitting, cracks for the final contour are created by blasting prior to the drilling of the remaining holes for the blast pattern. Presplitting works in loose as well as in hard rock. A typical example would be the removal of a 15-ft rock bench adjacent to a metropolitan avenue on which traffic must be maintained. The presplitting line holes could be spaced on 18 to 24-in. centers with a 2½-in. diameter. The upper 3 or 4 ft would be stemmed with dry sand, whereas the column charge might be made up of half sticks on 12-in. centers separated by dry sand. The bottom charge could contain three or four cartridges of 1½ × 8 Hercules Gelatin, extra 40 percent, for added strength in cleaning the bench toe. Consultants from the various powder manufacturers are available to make specific recommendations on strength, spacing, and burden for varying geological conditions.

5

General Procedures

5.1 Rock Anchoring and Bolting

Rock work often requires the use of artificial supports to maintain embankments and relieve stresses, for the assumption is that ultimate failure of ground is inevitable. Conventional supports, wood or steel, interfere with job progress because the project must be adapted around them. Rock bolts, which make ground self-supporting by consolidation, are embedded in unexcavated rock formations and thereby usually allow maximum flexibility in the project.

Types of anchorages common to building foundation construction are varied according to their function. Earth banks at the perimeter of a project can be stabilized by soldier beams and timber lagging. The soldier beams can be secured with walers and rakers, or rock anchors may be used. In order to be effective, rock anchors must be placed prior to the redistribution of stresses. The procedure is to drive soldier beams (12BP53) on 6-ft centers and then to place a 4-in. timber lagging between flanges. As the excavation proceeds, the anchors are placed and tested on a predetermined spacing of 5 to 10 ft or as designed (see Fig. 5.1).

A casing or open pipe (3-in. ϕ) can be driven through the overburden and reamed out. A rock drill is then inserted in the casing, and a socket is prepared in the rock formation to secure the anchor. In the case of a cohesive soil deposit, no casing is required because of the soil's natural ability to arch around the drilled opening. In the latter case, the grout pipe (2-in. ϕ) or casing is inserted in the drilled opening, and a rock anchor is then driven through the grout to the base of the socket. Quick-setting grouts are used so that the excavation can proceed without long delays.

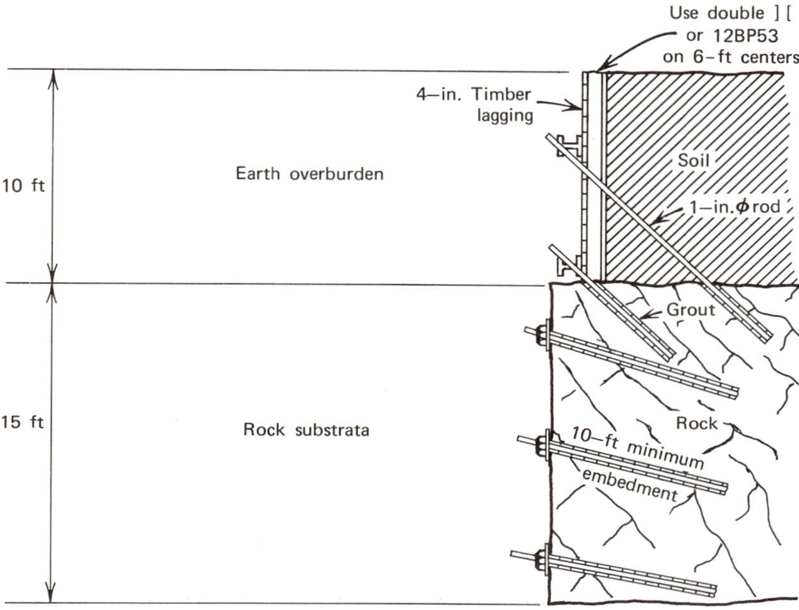

Fig. 5.1 Typical excavation illustrating the anchoring method of securing banks.

This procedure can be extended to secure banks of great depths (50 to 80 ft). In the foundation for the World Trade Center in New York City, basement walls are secured by anchorages into the bedrock of Manhattan. These "slurry" walls are up to 70-ft high and are secured by high-tension cables.

The depth of the socket depends on the characteristics of the rock and the load on each anchorage. In firm rock the necessary strength can be attained at a depth of 2 ft or less. Softer formations require greater penetration (Fig. 5.2 & 5.3).

All anchors should be preloaded to determine their effectiveness. It is practical to find the preload on a rock bolt in one of two ways. The first is to preload directly using hollow hydraulic jacks to a given pressure of 10,000 lb for a 1-in.-diameter bolt. The second and more common approach is to use a commercial-type torque-indicating wrench. This device permits direct reading of the torque applied to either the nut of a slotted-type bolt or the head of an expansion-type bolt. Tests using 1-in.-diameter slotted-type bolts showed that the torque-tension relationship was approximately 40 lb of load for each foot-pound of torque applied, provided reasonable precautions are taken to minimize undue frictional effects. The

Fig. 5.2 62nd Street and 3rd Avenue in Manhattan, 1965. Rock bolts pretensioned to a torque of 10,000 ft/lb secure a sliding formation that dips into the excavation.

Fig. 5.3 62nd Street and 3rd Avenue in Manhattan, 1965. Lagging between steel supports that have been tied back to a rock base has proved to be a safe and efficient support procedure.

accuracy of the relationship was ±27 percent. For example, a torque of 260 ft-lb applied to the nut produces a bolt load of 10,000 ±2700 lb 90 percent of the time. One-inch-diameter rock bolts are usually installed at a preload of approximately 20,000 lb.

Single-faced forms can be secured using rock bolts as anchorages for form ties. Walls that are secured in this manner help to stabilize rock formations by minimizing future weakening of the formation by seepage or creep. Vertical rock faces or slopes can be secured against movement by rock anchors. Chain-link fence fabric can be attached to prevent spalling rock from injuring personnel.

If prebid analysis or field conditions indicate that rock slopes may be unstable, then a rock-bolt system may be in order. A common method is to drill horizontal holes into the face of the excavation on 4- to 6-ft centers. Threaded rods are then grouted in place. Number 8 reinforcing rods are also often used. The advantage of grouting is that it provides a reliable binding through possible faults.

Cautionary controls should be instituted to check that the dowels or anchors are in fact placed horizontally. A highly inclined bolt may actually cause a rupture or slide.

Rod lengths should be standardized throughout an area with at least a 4-ft embedment into a sound anchorage.

5.2 Bracing

Bracing of rock excavations is necessary in areas where unstable slopes threaten to fail by creep or sudden movement.

Unsupported cuts, if not known to be self-supporting, should be braced during excavation. Stable rock slopes may not require any supports. A trained engineer experienced in geology should carefully inspect and recommend support procedures as the work progresses.

Tunnel roofs are held in place by timber or steel bents. These bents are placed while the drilling rig and drillers are preparing the next advance. The front section over the jumbo drill must be self-supporting during drilling and blasting. The timber or steel bent is erected and carefully blocked against the roof and walls; wedges are driven tight, thus prestressing the entire set. Lagging is laid across these caps to prevent falls of loose rock.

The bracing of a rock face assumes failure of the rock and resultant load transfer to the steel or timber support; whereas anchoring of rock actually strengthens the slope and thereby relieves the tensions that could accumulate at a failure plane.

The early support of a rock cut prevents serious stresses from developing into a failure plane. The bracing of the support for a retaining wall is shown in Fig. 5.4. For this project the rock was braced in sections, that is, the braces were set individually to assure a gradual transfer of stresses from the rock to the steel shores.

After the braces were secured and grouted to support the retaining wall, the toe or bottom of the slope for the wall was removed by explosives.

As evidence of actual load transfer, a careful record was kept of the wall location on completion of each of the following steps:

1. Excavate overburden rock.
2. Trench into rock bed at soldier beam locations.
3. Grout and secure soldier beams.
4. Remove toe of slope and prepare wall footing.
5. Pouring wall, thus encasing the soldier beams permanently as part of the basement walls.

On this job, a movement of 1-in. was recorded as the toe of the slope was excavated. This movement, as a result of the transfer of load from the footing to the lateral braces, was arrested. The serious requirement for bracing was known from geological data accumulated from experience and observation. Test holes and preliminary excavations indicated highly disturbed strata dipping at more than 60° into the cut. The existence of slickensides or probable failure planes was observed.

Faults perpendicular to the bedding planes are common and are due to the cooling and contraction of the earth's surface during periods of disturbance.

Figure 5.5 shows the installation of steel shores to assure the safety of the adjacent structure. The procedure was as follows:

1. Blast and remove the rock outside the influence sphere for the building footing. (A 45° slope is usually safe, and it can be increased as geological conditions permit.)
2. Line drill, blast, and remove a safe section along the wall footing. The figure indicates that a rock brace was installed on one side of the span before the berm rock was removed.
3. Install the concrete wall, filling all voids with concrete in the designated section.
4. Repeat the procedure above for intermediate sections.

The bracing of the rock, as indicated in step 2, before the berm slope is removed must be predetermined by an engineer qualified to judge geological conditions. The installation of the foundation walls as buttresses against rock pressure requires the design engineer's approval. The wall may have to be beefed up to support these added stresses.

Fig. 5.4 Ford Foundation building in New York City. A 25-ft retaining wall was braced prior to excavation. Soldier beams, encased as the new wall was poured, became a permanent part of Ford Foundation's basement walls.

Fig. 5.5 Shoring of adjacent building prior to excavation of rock slope. See Fig. 5.4 for expanded view of office building.

The method used to secure the forms can be extended to stabilize the rock face. Drill 1½-in.-ϕ holes into the rock face at a slight inclination to the horizontal (15°). These holes should extend into sound strata a minimum of 10 ft, uninterrupted by visible failure planes. Position cables, rock bolts, or threaded rods in the hole, and pour or pump grout into the opening, being sure that the grout is reaching the bottom of the hole. A grout tube extending the full length of the hole and withdrawn slowly often works well, once the workmen become experienced in its use. A quick-setting grout enables the rods to be tested by pretensioning to 10,000 lb after three days. The forms may then be set using the grouted rods as ties to secure the gang panel. The concrete is then poured, cured, and stripped; grouted rods are again tensioned to their design load of, for example, 10,000 lb. In this way, the rock does not have an opportunity to relax and transfer its stresses to possible failure plans.

5.3 Grouting

Rock formations can be reinforced by grout if there are voids, such as caverns or seams. The idea of grouting is to provide a fine material of high strength to fill the pores, seams, channels, or weak areas of a rock formation. Dams constructed on rock beds often require grouting to seal the natural channels between bedding planes and along failure planes. Tunnels constructed in rock formations may require grout screens to decrease the flow of water into the heading in advance of drilling.

Where it is necessary to grout formations containing channels of swiftly flowing water, cement grouts have not been successful, for the grout is washed away before it can set up. Chemical grouts may be helpful but should be used under the watchful eye of an expert, for these grouts, though useful under certain circumstances, may be harmful if used incorrectly.

Grout requirements vary with individual jobs, but generally they must be sufficiently fine-grained to pass into very small openings and cracks; they must be sufficiently fluid to permit pumping; they must form a substance that, when set, has adequate strength; and they must set with a minimum amount of shrinkage.

A cement-and-water mixture satisfies these requirements under normal conditions.

Grouting methods vary with the geological and field conditions of the project. A single-stage grouting operation can be used for shallow depths at very low pressures, but care must be taken to prevent raising the strata as pressure increases.

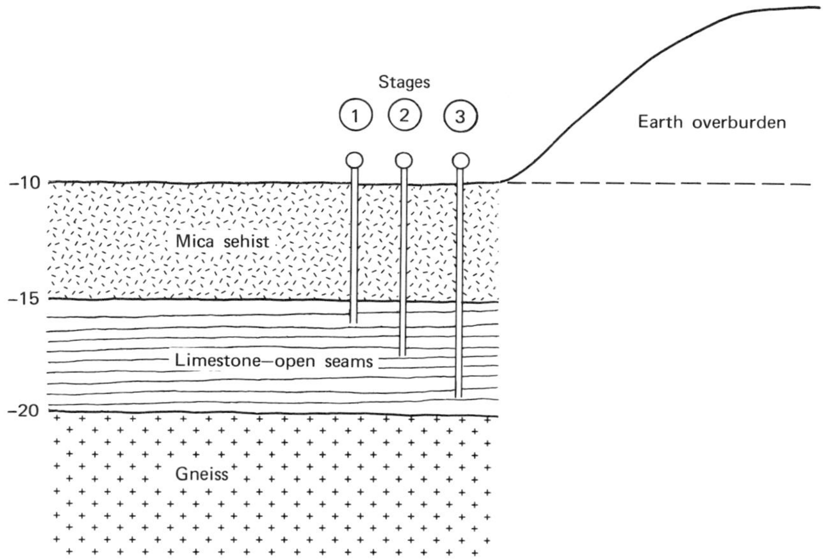

Fig. 5.6 Grouting of open seams in successive stages.

Successive-stage grouting is used in bedded strata where high pressures could open or lift the upper formation. The procedure is to grout the upper seams under low pressure and, by stages, grout the lower strata at increased pressures. This method consolidates the upper formation before an uplift pressure is applied (Fig. 5.6).

Grouting of bedrock may be required in foundation designs that superimpose high loads on concentrated areas of either porous or seamy rock. The problem with grouting is that it can often give the designer a false sense of security as to the soundness of the substrata after grouting is complete. Caverns and joints may remain unnoticed after the grouting operation has been finished. A test program can be called for where core borings are taken to substantiate foundation soundness.

5.4 Gunite

Certain types of rock have a tendency to absorb moisture and thereby lose their strength characteristics. Rock formations containing high concentrations of certain clay substances may actually expand laterally when exposed to moisture. Sealing the surface prevents further deterioration of the natural formations and thus assures the designer that the surface he inspects will be preserved.

Gunite is a fluid mixture of fine sand, cement, and water that is pumped through a nozzle. The spray of grout can be used to seal or protect the surface of a rock formation to prevent possible failure by sliding. During winter moisture may collect and form ice lenses that tend to heave the upper strata, thus causing probable failure. Sealing the surface with gunite, properly applied, prevents this.

Pneumatically applied concrete derives its name from the compressed-air transport of premixed dry cement and sand through a hose line of approximately 1½-in.-inside diameter. The materials travel at high velocities, varying from 100 fps (feet per second) at the hose-line entry to approximately 350 fps at the nozzle tip. Before leaving the nozzle, the mixture is instantaneously hydrated in a water chamber or body.

Pneumatically applied concrete is stronger, denser, has a higher specific gravity, has a lower water-absorption index, and generally is more resistant to weather and chemical exposure than conventional poured concrete. Gunite is excellent for sealing off rock faces.

European tunnel engineers have recently publicized this method, known as *shotcrete*, by using it under a variety of conditions. They have proved it to be a safe substitute for bracing under certain circumstances. Figure 5.7 illustrates the method under ideal conditions in a tunnel. The timing of the application of this method depends on the quality of the rock strata. If the strata are not self-supporting, steel or timber bents may have to be installed prior to applying the gunite. A wire or reinforcing cage may be installed between bents to stiffen the gunite lining. This procedure could be extended, if designed properly, to supersede the need for a lining.

The difficulty with the system is that as more procedural steps are added to refine and strengthen the application, the time span increases. Rock slopes or walls that are exposed tend to release their stresses, thus causing spalling and possible failure.

A middle road must be attained that protects the rock surface without interfering with the progress of the work.

5.5 *Groundwater—Effect on Weak Rock*

Foundation engineers spend a considerable portion of their lives searching for dewatering methods that will successfully overcome the problems presented by seepage through rock formations. Surface sumps continue to be the leading method for controlling this seepage (Fig. 5.8). The

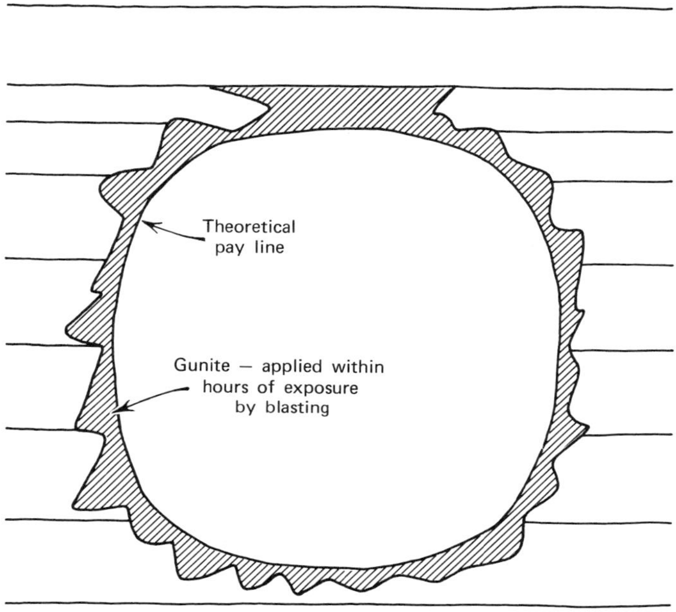

Fig. 5.7 Tunnel excavation. Gunite may be used as a protective barrier against stress release and possible failure.

prime obstacle on many jobs is to locate or construct a drain to carry water away from the job.

Many rock jobs are located in valleys, and the cost of transporting surface runoff and seepage away from the site is high. A storm drainage system that is convenient to the job is not always available, and the alternative is to discharge the water into a stream or river. Permission must be obtained from the proper authorities before expensive piping systems are installed. Periodic cleaning of discharge areas is always necessary. The deltalike deposits of silt and sand are an eyesore that may cause considerable opposition from public authorities.

Defects of a rock surface may be due to alteration or removal of rock constituents by chemical reaction between these constituents and water. Many rock constituents, such as feldspar, are gradually decomposed even by chemically pure water. Natural groundwater contains gases and solids in solution, which can considerably intensify the chemical effects of the water.

Water may enter the rock from the surface, or it may rise into the rock

Fig. 5.8 Ground water always collects in low areas, such as tunnels and footings. Air-sump pumps are normally used to remove this water.

from great depths. The main supply channels are joints and faults, and from them water enters the interstices between mineral grains in rock. Chemical alteration in fault zones can therefore be expected to be intense. Engineers and contractors should expect the worst in these areas until test pits or excavations indicate otherwise. The flow of water through a fault zone may be so intense that an excavation can easily be flooded in a short period of time. If underground channels are thought to exist, a testing and investigative program should be implemented to determine their extent. A grouting or guniting operation may be required to seal these channels before rock excavation begins.

The most commonly encountered soluble rock is limestone, which remains unaltered between channels, but day after day a certain amount of rock is carried away in solution by the water circulating in the joint system.

The water-caused internal decay of igneous rock, such as diabase, is associated with the formation of a greenish mineral known as serpentine. When serpentine is formed, a volume expansion occurs that almost always leads to exceptionally high static stresses in the rock. Excavations through this material release the internal stresses, and usually result in slaking at the sides of the excavation and gradual fragmentation of the bottom.

The porosity of dense rocks, such as granite, may be low; however, their permeability may be relatively high. The action of rainwater entering the rock is twofold: chemically unstable rock constituents, such as feldspar, are decomposed, and soluble constituents, such as calcium carbonate, are removed in solution. These processes cause a great portion of the earth's surface to be covered with a mantel of weathered or disintegrated rock (Fig. 5.9, 5.10).

Regions formerly covered with ice do not have this thick (100-ft) mantle of weathered rock because of the plow action of the moving ice sheet. The relatively recent ice passage has left little time for the development of more than a shallow (10-ft) mantle of soft disintegrated rock in the New York City area.

Metamorphic rocks, such as schists, consist chiefly of micaceous minerals, such as chlorite and mica. When exposed to the atmosphere, these rocks turn into a material hardly distinguishable from a clay. In some schists, the transformation seems to be due to a loss of cohesion and the mechanical break up of the original constituents of the rock. It is this claylike property that causes rock slopes to creep into the excavation. If the bentonite family of clays is present in the formation, movement may be the result of a swelling of the seams when exposed to the atmosphere. The bentonites have an affinity for water that constantly changes their volume.

Fig. 5.9 Rock excavation below river level can often test the truth of many theoretical solutions. Sheeting and shoring must be designed with the tides taken into consideration.

Fig. 5.10 Dewatering adjacent to the river was accomplished by means of (1) well points to top of rock combined with steel sheetpiling to enclose the bath tub, and (2) sump drains with chemical grout used to plug some of the larger channels.

Engineers should review the groundwater problem with caution and seek the advice of competent engineers familiar with the area before finalizing plans for construction. Constructors would be well advised to bid with caution jobs that are located in areas of intense weathering or areas having a history of underground drainage.

6

Special Problems

Landslides occur when the stress conditions necessary for failure in a rock slope are met. Landslides may involve materials of any kind, ranging from hard to soft rock or any combination of materials. A similar variety prevails among the processes that lead to slides. They include undercutting by erosion, man-made excavation, change in the groundwater levels, and progressive structural changes in the material adjoining the slopes.

Rock excavation means altering the stress conditions surrounding a project by either blasting or equipment (see Fig. 6.1). The alteration of natural conditions leads to a redistribution of stresses that can be satisfied only after a slope failure.

The most common causes of rock slides are the undercutting of a slope by excavation and the depositing of materials along the upper edge of the slope. Both operations produce an increase in the shearing stresses in the rock beneath the slope. Engineers should analyze geological conditions that may lead to a slope failure (see Fig. 6.5).

Rock slides can be analyzed on the basis of the "Mohr-Coulomb" criteria for failure as stated by the equation (see Fig. 6.3).

$$\tau = So + (p - hw) \tan \phi$$

Let τ = Average shearing resistance per unit of area of the surface of sliding.

So = Cohesion per unit of area.

p = Pressure per unit of area at a given point (A) of a potential surface of sliding due to the weight of the solids and the water located above the surface.

h = The unit weight of water.

w = The piezometric head at that point.

ϕ = The angle of sliding friction for the surface of sliding.

Fig. 6.1 Removal of the toe of steeply inclined strata must be supervised carefully. The apartment house above was not disturbed during the construction period.

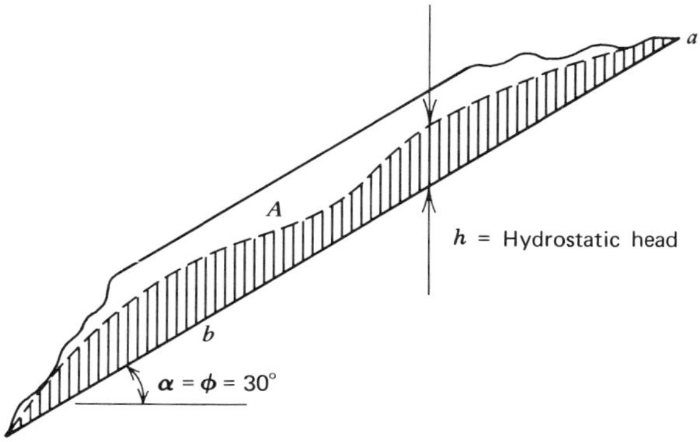

Fig. 6.2 Diagrammatic section through site of rock slide of Goldau, Switzerland, 1806, prior to slide. Upper slab was separated from its base by a thin layer of weathered rock. The dashed line represents the piezometric surface in this layer during a heavy rainstorm.

The action of the water pressure hw can be compared to that of a hydraulic jack; the greater hw, the greater is the part of the total overburden weight carried by the water, and as soon as hw becomes equal to p, the overburden "floats." If an increase in the piezometric head is coupled with a gradual decrease in shearing strength of the joint material, then a failure may result.

The classic example of this type of rock slide occurred in Goldan, Switzerland, on September 2, 1806. Figure 6.2 shows a section through the area prior to the slide. The upper slab, a conglomerate with a calcareous binder, is 5000 feet long, 1000 feet wide, and about 100-ft thick. During a heavy rainstorm, the slab moved down the slope, killing 457 people. This catastrophe was due to a gradual decrease in shearing strength caused by erosion and a rise in the piezometric surface.

A second example of this type of failure occurred at Italy's Vaiont Dam on October 9, 1963, when the worst dam disaster in history took place, taking almost 3000 lives. The catastrophe was caused by the recharging of the canyon banks with groundwater, which was sufficient to upset the stability of a delicately balanced steep rock slope as shown in Fig. 6.4. Note that the angle of internal friction (ϕ) was approximately 41° prior to the slide.

The trial set of computations for analyzing the stability of the slope is as follows:

$$\phi = 41^\circ \qquad \tan 41^\circ = 0.8700 \qquad \cos 41^\circ = 0.755$$

$$= 750 \text{ ft} \times \frac{160 \text{ pcf}}{144} = 833 \text{ psi}$$

$$= 833 \cos 41^\circ = 630 \text{ psi}$$

$$\tau = 1000 + 630\,(0.8700) = 1548 \text{ psi}$$

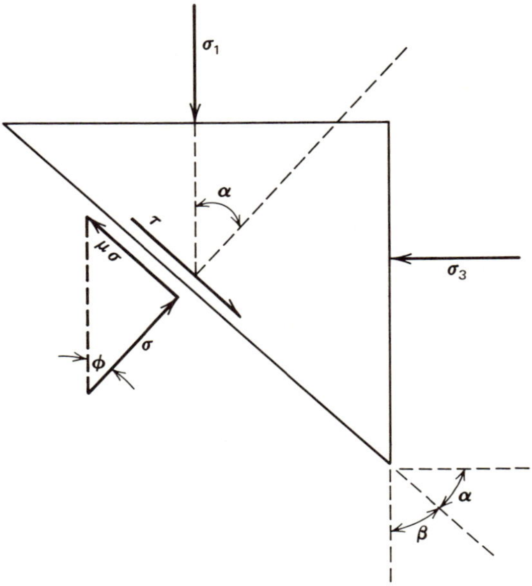

Fig. 6.3 Stress diagram of Mohr's theory.

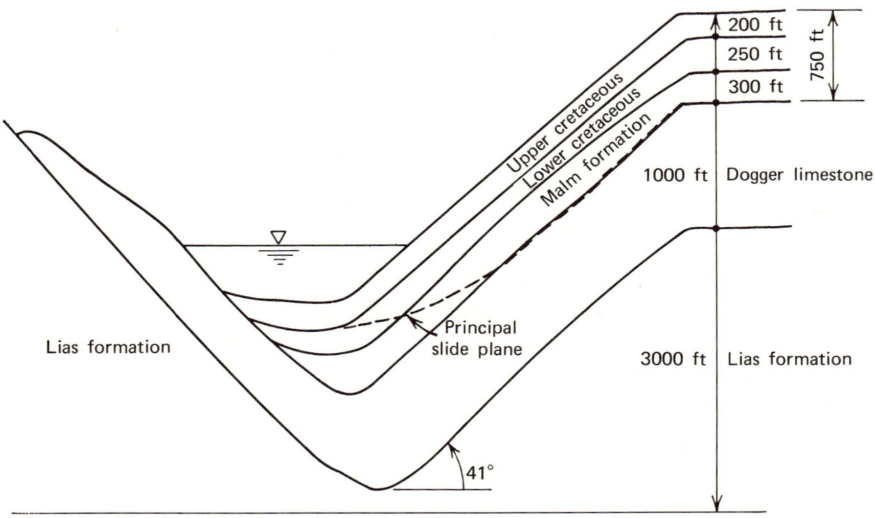

Fig. 6.4 Geological Cross Section, Vaiont Dam, Italy. Lias formation: thin beds of gray limestone alternating with thin beds of sandy limestone. Dogger formation: medium-bedded gray, dense-limestone. Malm formation: thin-bedded limestone with clay seams. Lower Cretaceous: thin beds of marl alternating with light red, thin beds of limestone.

Table 6.1 Rock Shear Strength

Type	Shear Strength (psi)
Sandstone, soft	1,500
Sandstone, medium	3,050
Sandstone, gray, hard	4,720
Sandstone, brown, fine-grained	3,600
Sandstone, gray friable, medium-grained	2,840
Limestone, fossiliferous, hard	4,160
Limestone, gray, hard	6,520
Limestone, medium, crystalline, very fossiliferous	7,600
Gneiss, gray, granitoid	9,300
Siltstone, sandy, medium hard	3,000
Dolomite	12,700
Quartzite	10,600
Clay seams[a]	10

[a] Foundation Engineering, Peck, Hanson, Thornburn.

Thus it can be seen, by comparison with the typical values given in Table 6.1 and by the knowledge that clay seams existed, that the shearing resistance developed in the slide area was close to the shearing strength of the material. In fact, the rise of the groundwater level caused the sliding forces to exceed the resisting forces as shown by the following computations: assume that the piezometric head was raised a total of 230 ft by the backing up of the water in the reservoir. Then

$$= 1000 + (630 - \frac{230 \times 62.5}{144}) \tan \phi$$
$$= 1000 + (630 - 100) \tan \phi$$
$$= 1000 + 460 = 1460 \text{ psi}$$

The shearing resistance is considerably less than the shearing forces, which were previously calculated, thus causing the slide.*

On a smaller scale, a similar analysis can be used to determine the forces present in all excavations that interrupt the natural bedding planes of a formation: an example follows (see Fig. 6.6):

$$W = \tfrac{1}{2}\,(21.8)\,(26)\,(170) = 48{,}100 \text{ lb}$$
$$N = W \cos = 48{,}100\,(0.642) = 31{,}000 \text{ lb}$$
$$F_1 = So + N \tan \phi \text{ (resistance to failure)}$$

The shearing resistance plus some outside stabilizing force have to equal the forces tending to slide the block into the excavation.

* George A. Kiersch, "Vaiont Reservoir Disaster," *Civil Engineering*, March 1964.

Fig. 6.5 Adjacent building columns may require braces before rock slopes are removed.

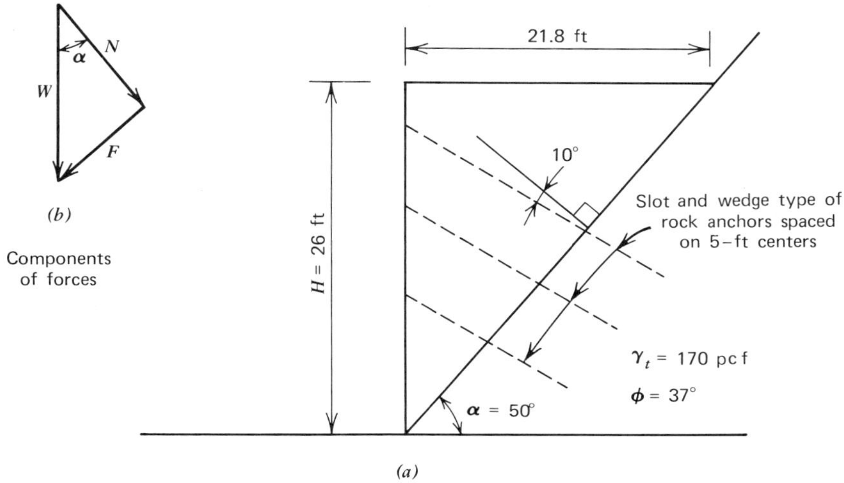

Fig. 6.6 Typical rock excavation.

$$
\begin{aligned}
F &= F_1 + Fo \\
N \tan &= So + N \tan \phi + Fo \\
31{,}000\ (1.19) &= 31{,}000\ (0.600) + Fo + So \\
Fo &= 18{,}400\ \#/\text{l.f. necessary to stabilize cut} \\
&\quad (\text{assuming } So \text{ is negligible})
\end{aligned}
$$

If the normal force is increased by using rock bolts or by some other artificial means, then the need for an external shore may be eliminated, as shown by the following example: assume that slot-and-wedge-type bolts have been installed as in Fig. 6.6, and that they have been stressed to a nominal load of 10,000-lb tension. These bolts are placed on 6-ft centers vertically, as shown by the dotted lines in Fig. 6.6, and are placed at an obliquity of 10° in a counterclockwise direction in order to impose both a normal force and a component force that resist the shearing force. Diagram and solution are shown in Fig. 6.7.

Thus it is seen that the Mohr-Coulomb failure criterion is useful and necessary in analyzing potential failure planes for open cuts, either natural or man-made. This same fundamental method of attack may be used in analyzing the forces at the tip of a pile or caisson founded on rock, the forces present in tunnel excavations, power house excavations, and at dam sites (see Fig. 6.8, 6.9).

In tunneling, horizontal and vertical pressures have always been of prime importance, resulting in empirical approaches that have been refined to a high degree of precision. The range of pressures to be expected

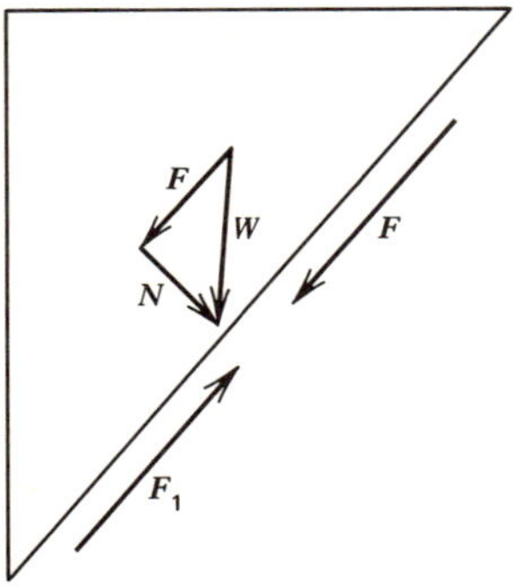

Fig. 6.7 $W = 48{,}000\#$; $N_1 = 31{,}000\# + 4(10{,}000)/5$; $N_1 = 31{,}000\# + 8000\#$; $F_1 = 375\#$ (up the plane); $F_1 = So + N_1 \tan \phi$. *For equilibrium.* $F = F + Fo$; $N \tan \alpha = So + N_1 \tan \phi + F_1 + Fo$; $31{,}000\ (1.19) = 39{,}000\ (0.600) + 375 + F_2 + So$; if $So = 13{,}225\# = 13{,}225/32.5 \times 1 = 408$ psf. Clay $Sa = 1440$ psf $= 10$ psi. Shoring would not be necessary.

in various zones of rock formations are shown in Table 6.2. These formulas, when applied to an imaginary tunnel through granite weighing 165 lb/ft^3 at a depth of 300 ft, give the results shown in Table 6.3, when the width B is assumed to be equal to the height, which equals 15 ft. If the granite has been chemically altered, squeezing and even swelling conditions may prevail. Loads may then attain the following values:

$$\text{Squeezing Hp max.} = 2.10\ (B + \text{Ht}) = 10{,}400 \text{ psf}$$
$$\text{Swelling } 20{,}000 \text{ to } 40{,}000 \text{ psf}$$

Thus the tunnel engineer has a range of values that can be used to help analyze the needs for support of a tunnel through rock of various characteristics. These values may be further analyzed by making use of the accompanying diagrams, Figs. 6.10 and 6.11. In Figs. 6.10, 6.11, if the spacing of the transverse joints is greater than the width of the tunnel, the rock layers act like solid slabs bridging the tunnel. The layers are then subject to bending stresses under their own weight. If these stresses exceed the tensile strength of the rock, then the roof must be supported or an arch will form. The probable maximum overbreak, if unsupported, is equal to 0.5 B. If the tunnel support is constructed and wedged soon after blasting, the friction forces on the sides of the rock fragments occupying the space between roof support and vault transfer part of the weight of this rock to the rock located beyond the sides of the vault. Hence even the ultimate load on an adequately constructed and backpacked roof support in closely-jointed, horizontally-stratified rock is likely to be much smaller than the value 0.5 B. On the other hand, if large empty spaces are left between the roof support and the roof, blocks will drop out of the roof, one by one, whereby the load on the support may increase to its maximum of 0.5 B.

If the strata are vertical, as shown in Fig. 6.12, then the individual blocks above the roof are held in place by the shearing strength of the

Fig. 6.8 Rock anchors grouted into the substrata may often be satisfactory for maintaining slope stability without braces.

Fig. 6.9 Rock tunnel under 42nd Street in New York City illustrates the irregularities to be expected in stratified rock. Note air mover for ventilation.

Table 6.2 Rock Excavation in Tunnels

	Vertical Roof Loads	
Description	Load	Remarks
1. Sound rock, hard and intact	0	Light lining, required only if spalling or popping occurs
2. Hard but stratified	0 to 0.5 B	Light support
3. Massive, moderately jointed	0 to 0.25 B	Load may change erratically from point to point
4. Moderately blocky and seamy	0.25 B to 0.35 $(B + \mathrm{Ht})$	No side pressure
5. Very blocky and seamy	0.35 to 1.10 $(B + \mathrm{Ht})$	Little or no side pressure
6. Decomposed rock, completely crushed but chemically intact	1.10 $(B + \mathrm{Ht})$	Considerable side pressure, softening effect of seepage towards bottom of tunnel; requires either continuous support for lower ends of ribs or circular ribs
7. Squeezing rock, moderate depth	(1.10 to 2.10) $(B + \mathrm{Ht})$	Heavy side pressure, invert struts are required. Circular ribs are recommended
8. Squeezing rock, great depth	(2.10 to 4.50) $(B + \mathrm{Ht})$	
9. Swelling rock	Up to 250 ft irrespective of value of $(B + \mathrm{Ht})$	Circular ribs required in extreme cases, use yielding support

Note. The tunnel roof is assumed to be located below the water table. If it is located permanently above the water table, the values given for types 4 through 6 can be reduced by 50 percent.

Table 6.3 Tunnel Design

	Roof Loads (psf)		
Rock Description	Minimum		Maximum
Spalling state—sound	0	to	400
Moderately jointed Hp max. = 0.25 B = 3.75 (165)	0	to	620
Moderately blocky and seamy Hp max. = 0.35 $(B + \mathrm{Ht})$	620	to	1730
Intensely shattered but chemically unaltered Hp max. = 1.10 $(B + \mathrm{Ht})$	1730	to	5440

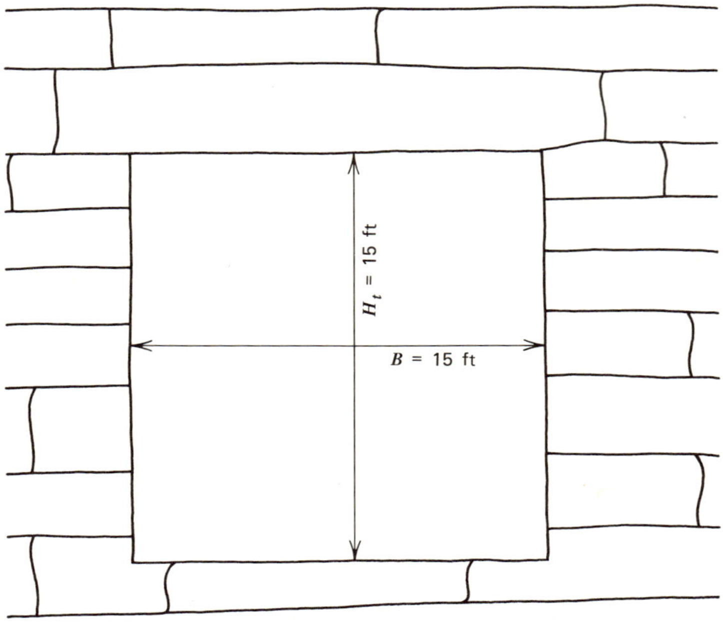

Fig. 6.10 Bridge action in rocks with widely spaced transverse joints.

irregularities between the two surfaces and the frictional resistance as demonstrated by the Mohr-Coulomb equation for equilibrium:

$$\text{(shearing resistance) } \tau = So + \tan \phi \text{ (repeat)}$$

In vertically stratified rock, the maximum probable overbreak when unsupported is 0.25 B. This should also be the design load for supports, unless there is reason to believe that the stratifications vary from this average markedly. It cannot be overemphasized that the key to minimum support loads is a well-backpacked support that is installed with a minimum of delay.

Figure 6.13 illustrates the side-pressure problem that might arise in steeply inclined strata. Force P must be capable of holding the wedge in equilibrium. Once again, the Mohr-Coulomb equation is a handy method of analyzing the forces that tend to cause failure (see Figs. 6.14, 6.15).

If the shearing strength of the joint-filling material is neglected, then a force polygon can be constructed to determine the magnitude of P and Q.

The side thrust may be estimated for design purposes using this procedure, always keeping in mind that a well-backpacked member prevents these maximum loads from developing.

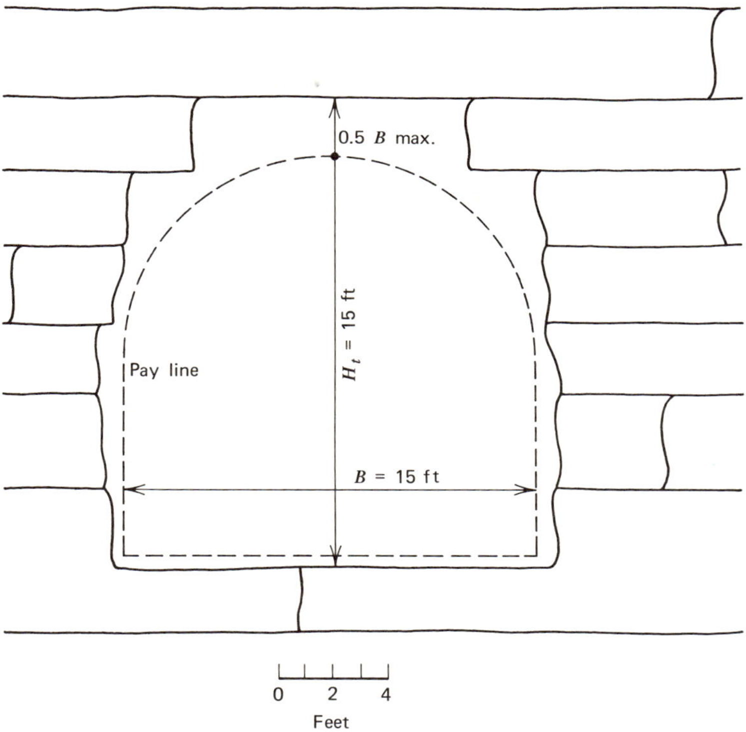

Fig. 6.11 Bridge action in rock with closely spaced transverse joints.

Many tunnels have been driven through decomposed rock, and it is interesting to point out the similarity between clay and decomposed rock formation. Some of these formations contain clay minerals of the Montmorillonite group, which gives them a tendency to swell under reduced load. This increase in the void ratio is due to an increase in the water content of the material adjacent to the excavation and a corresponding decrease in the water content of the material at some greater distance from the excavation.

A squeezing rock with moderate swelling characteristics may impose an initial load of 3500 psf, and after a few weeks this load may increase to 12,000 psf, which may cause local failures unless support has been provided. In deep tunnels through swelling rock, pressures of 20,000 psf are not uncommon. Exceptionally, pressures as high as 40,000 psf, have been encountered equivalent to the weight of a 270-ft-thick layer of rock.

T. S. Lovering published a graphic description of the behavior of such

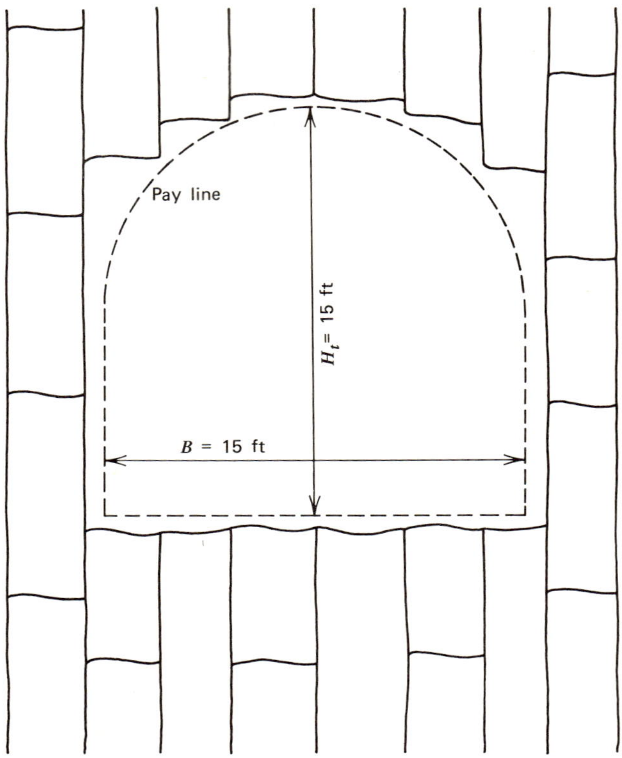

Fig. 6.12 Tunnel-through vertically stratified rock.

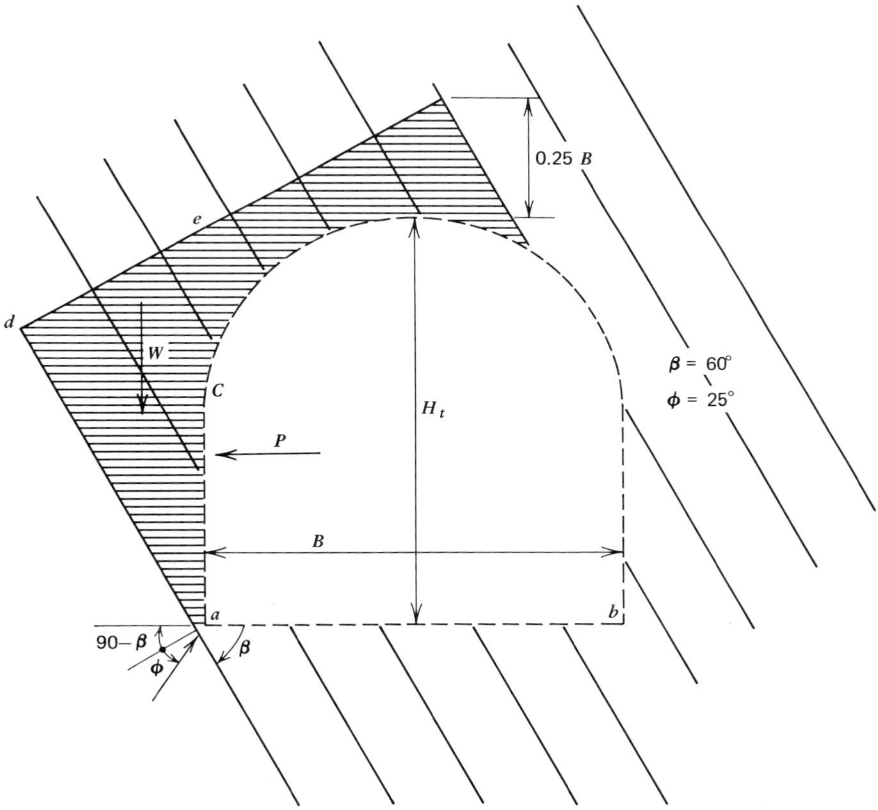

Fig. 6.13 Tunnel through steeply inclined rock strata.

Fig. 6.14 62nd Street and 3rd Avenue in Manhattan, 1965. Sliding rock can cause serious overbreak outside of the line holes. Steel rods can be grouted as shear pins through the rock strata, as seen above, to prevent slabs of rock from sliding into the excavation.

Fig. 6.15 62nd Street and 3rd Avenue in Manhattan, 1965. The presence of soft or weathered rock under adjacent structures may necessitate underpinning these structures to a firm foundation on sound rock.

rock, referring to a section of the Moffat Tunnel in which decomposed granite of the swelling type was encountered. "In this section, the ground swelled continuously and irresistibly. In the water tunnel, a bore parallel to the railroad tunnel, 12-by-8-inch timbers of Oregon fir were broken like matchsticks . . . On the eastern side of the fault zone, the walls were equally deceptive; very few gouge seams are present, and the rock seems to be bleached but sound granite. It is quite surprising to observe material of this appearance slowly closing up the tunnel without developing any cracks or shear planes. In fact, the behavior of the ground so impresses hard-rock miners as being uncanny that they will seldom work in this part of the tunnel for more than a few days before quitting."*

These materials can be handled only by empirical methods that make use of the limited theory available. With these questionable materials in mind, it is fitting that the civil engineer be responsible for the undertaking of practicable methods needed to overcome these obstacles.

* T. S. Lovering in *Rock Tunneling with Steel Supports,* edited by Proctor and White, published by Commercial Shearing and Stamping Company, Youngstown, Ohio.

7

Foundation Construction

General foundation construction covers a wide range of foundations, often differentiated by a variation in excavation methods.

7.1 Footing-Excavation Procedure

The simplest and most common type of design incorporates a variation of the spread-footing design (see Fig. 7.1). For this type of foundation construction the general area is excavated to subgrade (usually 1 ft below the bottom of slab) using explosives or other means. Reduced charges in the pier holes are supplemented by air tools and hand labor to complete the footing.

In New York City* and its environs the general practice under the code of the Department of Buildings is to have a representative of the department inspect the rock surface before pouring concrete (see Fig. 7.2). The inspector's function is to assure the public that the builder is conforming with city regulations regarding the seating of foundation piers (see Fig. 7.3).

The inspector ascertains whether in his opinion the rock foundation is satisfactory to support the design loads. The city code includes a table of allowable bearing pressures that the design engineer may use after determining the quality of the in-situ materials. The table is based on many years of experience with the particular geologic formation problems encountered in New York City.

A review of New York City's recent geologic history will help clarify the procedure for preparing footings. About 75,000 years ago, New York

* See the Appendix.

Fig. 7.1 General rock excavation is advanced prior to pursuing individual footings.

Fig. 7.2 Footings socketed into rock should be poured directly after inspection by the Building Department. Delays can promote surface deterioration.

Fig. 7.3 Pier holes should be completed rapidly, for the steel erector will want to set billets before mobilizing. (U.N. Delegates Building, East 47th Street and 1st Avenue, New York.)

City was covered by a 2000-ft-thick glacier that remained for more than 60,000 years. This latest of North American glacial ages is called the Wisconsin stage, and remnants of the glacier still exist in Greenland, Antarctica, and other areas.

As the glacier advanced at a rate of perhaps 1 ft/day, it stripped bedrock formations of their overburden and pushed jointed blocks of rock forward. Overburden removed exceeded 100 ft in low regions. This process exposed the older and stronger rock formations to a renewed cycle of weathering and decomposition.

As the glacier receded, it deposited its rock burden, leaving both stratified and unstratified accumulations. Plowed-up formations were deposited in deltas at the face, or terminus, of the glacier. Rough jointed blocks were ground smooth and further fragmented in this process.

The face of the glacier advanced and receded during each year, and this static equilibrium caused a buildup of unstratified and unsorted debris, or till, at the face of the glacier. A rock fragment plowed up by the glacier hundreds of miles north would thereby be transported and ground smooth by the creeping of the interior of the ice toward the face of the glacier.

A foundation engineer is thus presented with the problem of recognizing the state into which nature has chiseled the landscape surrounding the construction site. He must be alert to the limitations of preliminary investigations that seem to indicate a rock surface capable of sustaining building loads. Large deviations in the actual rock surface and the rock quality are common in any building area of an acre or more.

For this reason, pier footings that are to support substantial loads are pursued on an individual basis. The method used is to complete the general excavation and then lay out the column footings. If the rock surface is of suitable quality and the minimum depth (usually 3 ft) has not been achieved, then the superintendent must either blast the pier hole using explosives or chip the added depth using conventional equipment. A 3-ft or greater depth is usually handled by explosives.

Final preparation is always done by hand. This process requires that experienced pneumatic hammer men give the rock surface a neat and solid appearance. Footing surfaces must be stepped rather than sloped, soft corners or areas must be removed, and the footing surface must be cleaned and made ready for inspection and for concrete. The final operation for a foundation contractor is pouring the footing concrete. Many contractors have spent considerable monies on foundation work only to lose sight of the goal that the owner and designer are trying to achieve: a safe and secure foundation on which they may confidently build.

The fact that natural elements have alternately loaded and unloaded the landscape in 100-ton ft^2 increments helps the engineer to appreciate the variety certain to confront him in any excavation. The engineer is presented with a set of borings that generally indicate the depth and type of bedrock that is expected. The hardness of the material is interpreted as a function of the core recovery. That is, a recovery of greater than 75 percent in a 5-ft run usually indicates that the bedrock, at this level, is of sufficient strength and quality to meet the standards set for an allowable bearing of 40 tons ft^2. The engineer designs his footings on this basis, and then plans are distributed among interested contractors for bidding.

The contractor's engineers evaluate the borings and the design in accordance with their experience records for the area and prepare a bid that encompasses many value decisions as to probable depth of footings. In the New York City area it is common to request that the contractor take a "full gamble" on the quality of the rock and its depth when submitting a bid for foundation work. Experienced bidders know the variation in quality of the footing formation and therefore allow monies for the probable removal of extra materials because of the unlikelihood of finding good bearing material at minimum depths. The recommended procedure is to request bids for minimum depths of foundations and provide unit prices for additional work in increasing the footing depth to find a quality bedrock that meets the design requirements.

Either of these procedures is acceptable if both the designer-owner and the contractor are aware of the risks and make suitable allowances for them. However, a major dilemma could arise if the designer optimistically used an allowable bearing pressure that was greater than practical, and if the contractor unwittingly guaranteed an impossible dream. The contractor in search of an acceptable foundation continues to excavate, ever increasing the depth of the pier holes. The designer, convinced that his earlier design decision of a particular allowable bearing is practical, holds that the contractor must continue to remove the overburden to deepen the pier hole until suitable bearing is found.

A contractor faced with this dilemma must submit to legal counsel and discontinue his search until a practical redesign is provided. A pier-hole group whose depth measured from subgrade is anticipated to be greater than 15 ft when its minimum depth is 3 ft should be cause to alert the owner-designer team that a redesign is required. The basis for this recommendation is that pier holes whose depth exceeds 15 ft are unwieldy and inefficient for standard methods.

A change in bearing design from 40 to 8 tons ft^2 usually removes all obstacles to a speedy foundation completion. An example is as follows:

Building area	200×200 ft	40,000 ft^2
Column grid	20×20 ft	400 ft^2
Column required	=	100
Column loading $P =$	=	800^k
40-ton/ft^2 footing size	$\dfrac{800 \times 1000 \text{ lbs}}{40 \times 2000 \text{ lbs/ft}^2}$	10 ft^2
8-ton/ft^2 footing size	$\dfrac{800 \times 1000 \text{ lbs}}{8 \times 2000 \text{ lbs/ft}^2}$	50 ft^2

The result of the illustrated calculations is to show that by increasing the size of the footing from 3.3×3.3 to 5×10 ft, the allowable bearing may be decreased from 40 to 8 tons/ft^2. If it is impractical to seek a high bearing, then the design-contract team should work together to arrive at an acceptable reduced value that will meet the standards of the local building department.

This procedure has been outlined for a classic condition where the surface has been found high above the surrounding water table, and the chief decision is concerned with the proper bearing value to be assigned to the footing.

7.2 *Drilled-in Caissons*

A more difficult design is approached when a building site contains unsuitable bearing material overlying the rock surface and a high water table (see Fig. 7.4). The cost of excavating a pier hole under these conditions becomes prohibitive because extreme measures, such as compressed air, might be required to complete each shaft or pier hole (see Fig. 7.5). A comparatively recent solution to this type of design problem is the use of "drilled-in caissons." These composite columns consist of a heavy wall pipe, concrete filling, and, depending on the design load, a wide flange, jumbo, or built-up steel-core section terminating in a drilled socket through which the load is transferred by direct bearing and bond to the rock (see Fig. 7.6).

The procedure to complete a caisson is listed below:

1. The shell is driven or drilled to bedrock. Shells vary in size from 12 to 42 in. in diameter.
2. The shell is cleaned by augering or clamshelling.
3. A heavy-duty percussion drill bit is inserted in the shell and a socket is cut.
4. The steel core of the desired length is ordered and cut in the mill.

Fig. 7.4 140 Broadway in Manhattan. Joint venture by Thomas Crimmins Construction Company and Spencer, White, & Prentis. Excavating to subgrade can be difficult when working between previously installed 42-in. caisson shells.

Fig. 7.5 140 Broadway in Manhattan. Joint venture by Thomas Crimmins Construction Company and Spencer, White, and Prentis. The socket in large-diameter caissons may have to be cleaned manually to assure proper seating.

Fig. 7.6 140 Broadway in Manhattan. Joint-venture. Caisson caps must be secured to the jumbo steel core by means of shear connectors and reinforcing.

The crew lowers the core into the shell and grouts the base into the rock.

5. The entire shell is then concreted.

Drilled-in caissons were developed by J. H. Thornley in the early 1930s, and were patented around 1935. Early applications were in Cincinnati and on some Cleveland bridge foundations. The installation procedure is an adaptation of techniques for drilling deep oil wells. In 1936 a test was conducted in New York City on a 12-in. drilled-in caisson; it carried 414 tons with a settlement of 0.8 in.

A committee of experienced engineers was appointed for further study. The committee investigated a number of projects where this relatively new support system was in effect. One project, the Holland Tunnel ventilation shaft, had 24-in.-diameter caissons socketed into the bedrock at a depth of 180 ft. Another project, a building within the Brooklyn Navy Yard, had 30-in.-diameter pipe seated 157 ft below subgrade. These projects led the committee to conclude that buildings constructed on long-pipe piles or drilled-in caisson foundations would experience negligible settlements. The committee recommended acceptance of this new foundation procedure.

The bond between rock and concrete was tested in tension to 386 psi without any sign of failure. An appointed committee of the American Society of Civil Engineers evaluated this new concept, verifying the tests and visiting two bridge foundation sites in Cleveland to inspect the entire operation and an actual finished socket. The committee investigated the slenderness ratio and lateral restraint of the soil, especially for long (100 to 200 ft) lengths.

Settlement records on long-pipe and drilled-in caisson foundations for the Holland Tunnel ventilation shaft (180 ft/24-in. ϕ pipe) and the Brooklyn Navy Yard building (77 to 125 to 157 ft/30-in. ϕ drilled-in caissons) showed negligible settlements, and the committee concluded that slenderness ratio/lateral restraint was not a significant design criterion.

Approval for use in New York City was filed before the Bureau of Standards and Appeals in January 1939, with approval given in December 1942 (see Table 7.1).

The recently revised New York City building code defines caisson piles (more properly, drilled-in caissons) as concrete-filled pipe piles that are socketed into bedrock of Class 1–65 (hard sound rock—60 ton/ft^2), 2–65 (medium hard rock—40 ton/ft^2), or 3–65 (intermediate rock—20 ton/ft^2), and constructed with steel cores. Specific details cited in the new code cover items previously stated in the Board of Standards and Appeals approval updating the material stresses; they allow for increased rock-bearing pressures due to embedment and spell out seating of the shell as

Table 7.1 New York City Building Code—1968

Basic Allowable Bearing Pressure (ton/ft^2)	Description	Characteristics	Depth of Embedment
60	Hard, sound rock (Fordham gneiss, Ravenswood gneiss, Palisades diabase, Manhattan schist)	Rings when struck, will not disintegrate after exposure, unweathered cracks with an 85% core recovery in a 5-ft run (cracks—⅛-in. wide at spacing 3 ft)	Increase by 0.1 × basic value for each foot of additional embedment but not exceeding 2× basic value; rock to be sound and not shattered by blasting
40	Medium hard rock (includes above plus Inwood marble and serpentine)	Same as above except cracks may be slightly weathered—¼-in. wide at spacing 2 ft, core recovery—50% for 5 ft run	Same as above
20	Intermediate rock (includes above plus cemented shales and Newark sandstone)	Same as above except dull sound when struck, fractured and weathered zones—1-in. wide at spacing 1 ft, core recovery—35% for 5-ft run	Same as above
8	Soft rock (includes above in partially weathered condition plus uncemented shales and sandstones)	May soften on exposure, weathered zones, 3-in. wide and filled with stiff soil, core recovery—35% for 5-ft run, *N* value—50 blow/ft; if core recovery—20%, need 3-in. ϕ cores for soil classification	No provision in code

sufficient to stop the inflow of soil. The design-decision factors in the code state that the allowable axial load on a pile will be the least value permitted by consideration of the following factors:

1. Capacity as a structural member.
2. Allowable bearing pressure on strata underlying the caisson tip.
3. Capacity as indicated by load test where load tests are required.

Thornley has stated that the caisson is primarily intended to meet the really tough conditions where loads are high, rock bearing is desirable, and the overburden difficult to penetrate. The load-carrying capacity has been increased dramatically from a 414-ton load test on a 12-in. drilled-in caisson in 1936 to current design loads of 3700 tons on a 42-in. drilled-in caisson.

For the combination of high load (greater than 1000 tons) and great depths below subgrade to rock (greater than 100 ft), the drilled-in caisson probably represents the most economical foundation process. The relatively high cost of seating the cutting edge of the shell and drilling the socket becomes a fixed cost, tending to make the drilled-in caisson less expensive for greater depths than comparable load-carrying processes for the same great depths.

On the other hand, with rock at a shallower depth below subgrade (25 to 50 ft), unless the loads are very high, the fixed cost of seating the cutting edge of the shell and drilling the socket becomes a major portion of the cost, and nondisplacement piles bearing on rock might be less expensive. Pier holes to rock at these shallower depths also might compete in economy with drilled-in caissons in certain soils.

The assumption of Class 1–65 or 2–65 (hard-sound or medium-hard) rock for socket design gives a flexibility in design for the actual rock found at the construction site. That is, the socket depth can be adjusted to conform to the actual quality or class determination of the socket rock as determined from the borings. It might be advisable to take the precaution of extra borings, perhaps larger-diameter cores and even testing of cores where rock of Class 3–65 (intermediate rock—20 ton/ft^2) is contemplated for drilled-in caisson sockets. A possible reduction of design bond stress might be indicated; bond stress might require laboratory or a field test program. Full respect for the design loads and quality of rock socket is necessary.

Controls and inspection concepts are important to the in-situ performance of the foundation. What you do not see may hurt you and/or the foundation being constructed. Because of the high loads involved in drilled-in caissons, the as-driven location and plumbness control are extremely important. Location at cutoff is important for redesign consideration as soon as practical and before the core is set. The following corrective measures might be taken:

1. If there is room inside the shell the core can be set closer to or on the proper location.
2. The core can be rotated to provide greater resistance to eccentricities.
3. The core can be increased in size by the addition of plate on web and/or flanges.
4. The socket can be lengthened.
5. Combinations of the above may be required.

The seating of the cutting edge of the shell in rock "sufficiently to stop the inflow of soil" is the key operation prior to drilling the socket. Where water conditions and the seating of the cutting edge in rock permit visual inspection of the drilled socket, there can be greater correlation between other sockets, drilling times, and boring information. The engineering data sheet required for Board of Standards and Appeals approval is an important record summary of installation.

The general rule for any type of rock foundation is to level and clean the surface prior to pouring the concrete foundation. The designer and contractor must remember that many rock surfaces deteriorate on exposure to atmospheric conditions. Therefore, it is imperative that wall and columns be sealed immediately after inspection.

8

Operating a Construction Business

8.1 Structure of the Organization

Organizing and managing a construction business differs only in detail from managing any other profit-making enterprise. In order for a contractor to realize a profit, he must plan his operation carefully so that he can control those facets of the business which determine the profitability of the firm's undertakings (see Fig. 8.1).

The fundamental divisions of a construction business are executive, estimating, contracts, personnel, engineering, purchasing, accounting, payroll, and field construction. They must function not only within the contractor's organization, but in the interwoven relationships with architects, engineers, general contractors, subcontractors, manufacturers, material dealers, equipment distributors, labor, and, finally, the general public. The contractor must be prepared to serve; he and his personnel must be willing to sacrifice their individual desires to that of the client, whether public or private. An organization that ignores the social developments of our nation is of little value to society, and usually attracts selfish individuals who tend to disrupt rather than unify organizations.

The contract documents generally outline the requirements of a given project, including progress schedules, submission of shop drawings, and the employment of a competent superintendent. In addition to the requirements set forth in the contract, it is presumed that the contractor has an organization that will handle the project in a businesslike manner.

An owner or client expects that his purchase of a complete foundation in rock from a qualified bidder will preclude his having to perform any of the duties or responsibilities he has included in the contract. The key, for an owner, is to predetermine the qualifications of the contractor before

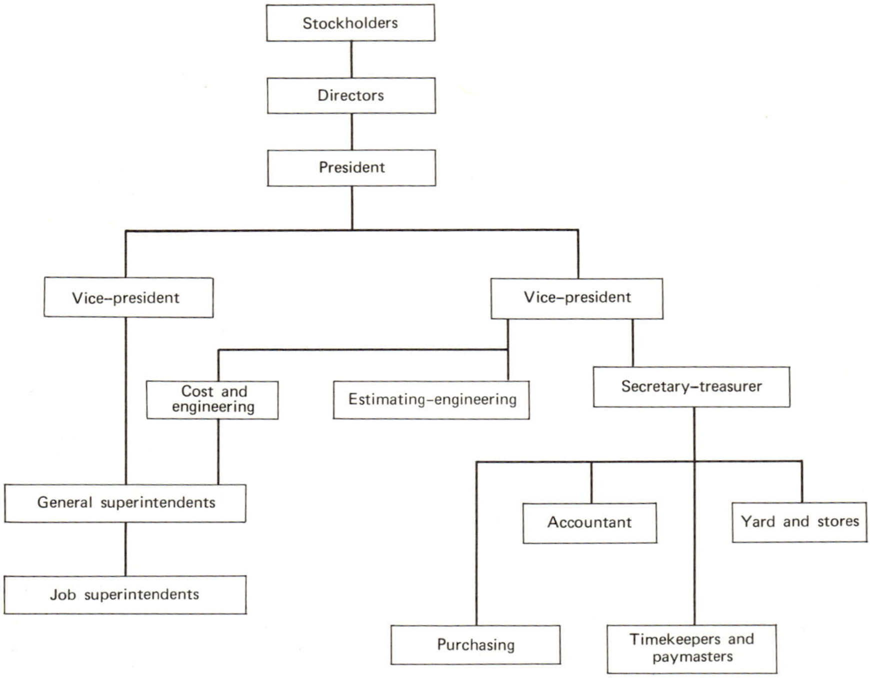

Fig. 8.1 Organization chart for a contracting corporation.

requesting a bid. This can be done by checking his experience and by requesting a performance bond guaranteeing completion of the contract.

The experience of a contractor is extremely important. Contractors have a fatality rate in business of more than 75 percent. Granting a contract strictly on the basis of the lowest bid may be an invitation to legal proceedings, for if the contractor is not qualified, he may go bankrupt.

The qualifications for a business enterprise are similar whether the business is contracting or manufacturing. A business enterprise must be based on the organization of an executive or owner whose responsibility is to delegate authority to his subordinates in a manner that contributes to a profitable venture. He must be familiar with the general operation of all departments, and have reports submitted for his consideration in order to review periodically the efficiency of each department. Departments that are not functioning efficiently must be updated by new methods or by changes in the personnel. The executive should always remain in the position of evaluator; he must be free to consider major decisions that affect the business.

In the beginning of any venture the principals must decide on the type of organization that will best suit their needs. The sole proprietorship has the greatest freedom of action, but its capacity for growth is limited to the ability and energy of the owner, and its profits are taxable as personal income. The partnership offers greater depth in management capacity and better credit possibilities, but the life of the partnership can be ended by the withdrawal of a single partner.

The corporation permits limitation of liability, and is explored in depth as a model for most small contractors interested in rock excavation. The model corporation shown has a capability of performing in a gross volume range of from $1 to $20 million/year. This range presumes that approximately 75 percent of this volume will be performed by the contractor's own forces.

Stockholders own the business. They elect directors and delegate to them full authority to manage the business. The directors function as a group and must formulate policy regarding types of construction, financial planning, and long-term contracts. They should review the corporation's operating policies and financial standing. The chairman of the board speaks for the group. The model corporation has a vice-president in charge of field operations. General superintendents report to him the progress of individual jobs.

The vice-president of the model corporation assumes responsibility for acquiring contracts with desirable clients in work classifications that conform to corporation policy. He is also responsible for the establishment of operating policies, for the implementation and enforcement of approved policies, for the coordination of all functions of the business, and for the overall profitability of the business.

The vice-presidents, elected by the board of directors, are responsible to the president for planning and directing the overall functions of engineering, estimating, and bidding; coordinating field operations with the administration, particularly with respect to scheduling, purchasing, contracts, subcontracts, and cost accounting; and obtaining all possible tabulations of bids in the corporation's area of concern, analyzing them for information useful in contract negotiations and bidding.

The secretary-treasurer, although elected by the board of directors, is responsible to the president for planning and directing the overall operation of the purchasing, accounting, timekeeping-payroll, and office services; administering all payments, collections, and credit; and preparing and filing all tax returns.

Every business finds it advantageous to vary its organizational structure not only to meet the particular needs of the corporation, but also to utilize the particular characteristics of its officers. However, a formalized corporate structure is necessary so management may be sure that all the business functions are being evaluated and continually kept efficient. Frequent informal meetings or luncheons are an asset to a corporation because they open the channels of communication on which every business depends.

8.2 Cost Methods

A construction firm is expected to predict the cost of a given venture prior to the award of contract. If the prediction is accurate, the company stands to gain a fair return on its investment. If the estimate is inaccurate on the negative side, the company may go bankrupt. If the work is overestimated, the company will probably not get the job, as the low bidder *may* have estimated the cost more closely.

Accurate estimates are clearly worth striving for, and may be attained only if actual costs are kept in a manner that lends to simple interpretation (see Fig. 8.2). After a contract is negotiated, the engineer should prepare a cost code that both the field and the office personnel will understand. For example,

Code	*Description*
101	General earth excavation (including berms)
102	General rock excavation (including berms)
103	Masonry removal
104	Column, wall, trench, and pit excavation, including grading and hand leveling—earth
105	Column, wall, trench, and pit excavation including grading and hand leveling—rock
106	Line drilling
108	Pumping
109	Backfill
201	Sheeting—timber—soldier beams, including bracing
202	Underpinning
203	Sheeting—column footing
204	Steel vertical sheeting, including bracing
205	Timber ramps
300	Piling
400	Steel framing
501	Pile caps—forms, concrete
502	Construction fence
503	Column and wall footings—forms, concrete
504	Grout billets, grillages
505	Curing and protection
506	Reinforcing
507	Encase grillages
508	Wall forms, concrete (including pits)
601	Finish surfaces
701	Supervision

DATE 3-6-71

JOB AMERICAN AIRLINES

FORMN. RUFO

NAME	CL.	GEN'L. EARTH	PIER ROCK	MASONRY REMOVAL	LINE DRILLING	PUMPING	UNDERPINNING	SHEETING		
ROGERS		2		2		2		2		
KRAMER			8							
DOUGLASS					8					
FINN							8			
MAREK			2			6				
GALE				6				2		
COAKLEY		6			2					
SOKOLITCH			4							
COLLINS				4						
WALLE		4					4			
WALSH			2	2		4				
HOPE				4				4		
REID		2			4		2			
LARSEN			2		2			4		
DILLON				2		2		4		
TOTAL		14	18	20	16	14	14	16		

Fig. 8.2 Foremans work card—man-hours are distributed over cost items.

The cost code provides a convenient procedure for charging all work costs and for describing daily activities. The job superintendent or engineer should denote on a daily basis those items in the code worked on by the tradesman. A sample daily report follows for easy reference (Fig. 8.3).

The job diary continues to be the lifeline for a corporation because it reflects the efficiency of the project. Its purpose is to provide a breakdown of costs for comparison with the estimate, and to give a record of job progress by both the contractor's and subcontractor's forces. Diaries are considered in courts to be substantial and accurate accounts of job happenings.

The labor and equipment breakdown into a cost code coupled with the quantities completed combine to present a strong picture for analysis by management. The record of subcontractor progress and any delays or accidents tends to fill out the picture of overall job progress on a specific date.

Invoices for materials and equipment rental are coded in a similar fashion by the job superintendent, or his assistant, who should be responsible for the propriety of the billing as far as work performed and material received. This combination of invoiced costs and direct labor, when added to subcontractor obligations, determines the actual cost of an item of work, and this may be compared with the estimate.

A continuing comparison of bid costs versus actual costs is necessary for the overall efficiency of a corporation (see Fig. 8.4). A business that considers cost comparisons wasteful has taken its final step toward self-annihilation. No company can survive on guesswork or wishful thinking.

Field personnel who predict costs on the basis of their past experience tend to underestimate actual costs. The reason is simply that human beings are optimistic and tend to remember only the periods of past jobs that functioned as planned. Usually they neglect such costs as insurance, shop stewards, delays due to breakdowns, and other items of cost. Spot cost checks are helpful during a job, but should be tempered with the knowledge that they may be 10 percent below actual costs because of optimistic reporting of an average day's production without considering rainy days and nonproductive labor.

An organization that has set up an effective cost control system, and is ready to communicate this information on a regular basis to the estimating department, is prepared to perform its contracts at a profit. Costs may be lowered by cutting corners and decreasing safety standards, but this is not the intent of a cost control system. It is the function of the executive in charge of field operations to see that all work performed meets the corporation's standards for excellence and quality. A safety program that keeps the job superintendents aware of this area of their responsibility may be worked out with the insurance carrier. Safety is good business, for a

Report No. 62

THOMAS CRIMMINS CONTRACTING COMPANY
41 EAST 42nd STREET, NEW YORK, N. Y. 10017
TEL: 986-3335

SUPERINTENDENT'S DAILY REPORT

Contract G 8 Location AMERICAN AIRLINES Date MAR. 6 1971

Weather FAIR 36°

NO. MEN	CLASSIFICATION	HOURS	LOCATION AND DESCRIPTION OF WORK, INCLUDING OUTPUT
1	JOB SUPT., PROJ. ENG., TIMEKEEPER	8 hrs EA.	SUPERVISION
1	SURVEY CREW - Chief, transit man, Rodman	8 hrs EA.	SURVEY
1	DOCK FORE	8	UNDERPINNING Adj. building NORTH OF SITE (4/16) SHEETING SOUTH SIDE (4/16)
4	DOCK BUILDERS	32	
1	BLASTER	8	ROCK EXCAV. (PIERS A5, A7, A9) 8
1	POWDER CARRIER	8	" " " " 8
6	DRILLERS	48	LEVEL FTGS. A6, A7, A8 24 hrs LINE DRILL EAST BANK 24 hrs
1	EXCAV. FORE.	8	GEN'L EARTH (14), PIER ROCK (18), REM. MAS. (2/20),
15	EXCAV. LAB.	120	LINE DRILLING (16), PUMPING (14), UNDERPINNING (14), SHEETING (16)

DELAYS, CHANGE ORDERS, ACCIDENTS & VISITORS	EQUIPMENT ON JOB
DELAY - H. JACKSON ORDERED US TO STOP ALL WORK NORTH OF H LINE WHILE ENGINEER REDESIGNS AREA. (WRITTEN CONFIRMATION HAS BEEN REQUESTED)	1 - SHOVEL - 2 c.y. LIMA
C.O. #7 - RECEIVED - AUTHORIZING EXTRA WORK ALONG STREET SOUTH OF LOT.	11 - TRUCKS - BREEN
ACCIDENTS - 62 days - NONE	4 - WAGON DRILLS
VISITORS: A.H. CRIMMINS 10 AM A. GORDON 11 AM	3 - 1200 COMPRESSORS

Signed Bernard Monahan

Fig. 8.3 Field Report.

THOMAS CRIMMINS CONTRACTING COMPANY

JOB: FORD FOUNDATION
321 EAST 42ND ST.

WEEK ENDING: 6-1-65

ITEM: LINE DRILLING

QUANTITY		UNIT COST		TOTAL COST	
FIGURED	ACTUAL	FIGURED	ACTUAL	FIGURED	ACTUAL
47,400 l.f.	47,000 l.f.	0.60¢/l.f.	0.54¢/l.f.	$28,440.–	$25,260

BASIC UNITS			UNIT AVERAGES
No.	JOB CLASSIFICATION	RATE	
	DRILL FOREMAN 100 hrs		
	DRILLERS 400 hrs		
	FORE. LAB. 150 hrs		
	LABORERS 700 hrs		
			$11,000
	33% INS. & WELF.		3660
	TRUCKING & EQUIP.		8700
	MATERIALS		1300
	YARD		600
			25,260

REMARKS: INTERIOR LINE HOLES WERE PLACED ON 2' CENTERS AND LIGHTLY LOADED TO PRESPLIT PERIMETER. WORKED WELL.

HARLIN 1730

Fig. 8.4 Cost Report.

good record provides bonuses in the way of dividends and lower premiums than paid by competitors.

In estimating new work, the following items must be determined:

1. The quantities to be excavated or to be installed, the delivered price of materials needed, and the time required for manufacturing.
2. The cost and extent of temporary facilities.
3. The portions of the work that could be handled better by a subcontractor.
4. The predicted progress schedule and the cost of possible penalties.
5. The costs of bonds, insurance, loans, licenses, and fees.

Estimates should be prepared accurately and in detail so that construction cost accounting and progress reports can be expressed in terms of the estimate for comparison purposes, and so that field supervision will know what methods the estimator had in mind when he arrived at the prices bid.

The owner or general contractor usually advertises that plans are available and the date on which estimates are to be submitted. An engineer, having determined the desirability of the work, then sets out to make a quantity survey of the drawings. Accuracy and consistency are the key actions to successful bidding. A job engineer who measures actual rock excavation, for example, in pier holes, and then neglects to compare it with the estimate, which may have allowed only 6 in. on each side for overbreak, tends to confuse and disrupt an organization's harmony and possibly its profit margin on future contracts. Pier or trench excavation can be computed as desired, but a comparison of actual and estimated quantities as well as costs must always be presented.

Information helpful in analyzing the difficulty in fulfilling the contract should also be collected and summarized. Old maps or plot surveys found in city or county planning departments are always helpful. In New York City each borough's Hall of Records keeps an accurate file on each land parcel and any alterations or improvements to it. A geological map showing the type and quality of rock indicated by borings is also helpful.

This information is valuable in the final phase, which is to apply predicted unit costs to the quantities that are known to be included. The markup, or profit, is then added to the subtotal to complete the bid. Bid openings are tense melodramas in which the low bidder cannot help but wonder if he has left something out of his estimate. A contractor may hope to be awarded anywhere from 5 to 10 percent of the jobs he bids. Every experienced contractor knows what it means to lose money on a contract; the goal is to minimize losses and maximize profits. Average 1971 unit costs are listed on the following page:

	Quantity	Unit	Cost
Fence—8-ft Plywood	500 lin ft	$ 8.00	$ 4,000
Earth excavation	2,000 yd^3	$ 4.00	$ 8,000
Masonry removal	200 yd^3	$40.00	$ 8,000
Rock excavation	15,000 yd^3	$10.00	$150,000
Line drilling	50,000 lin ft	$ 0.60	$ 30,000
Timber sheeting	3,000 ft^2	$ 5.00	$ 15,000

8.3 Equipment Records

Equipment records are kept for the purpose of analyzing the cost of operating a piece of equipment versus the standard rental rate charged by suppliers. For each job a monthly equipment report should be submitted that indicates the days worked by company-owned equipment. This report is then used to summarize the activity of a piece of equipment, including breakdowns, during a given month. Depreciation, repairs, and other costs are computed and compared with the cost of rental. A yearly summary evens out the peaks and valleys, and clarifies the value of ownership.

Equipment ownership can be deceiving, for a service yard must be maintained for storage and repairs. Many contractors have failed because they naively overcapitalized their store of accessory equipment. Materials stored in a contractor's yard have to be rehandled on a regular basis, and the cost must be prorated. In effect, the contractor finds himself involved in a service business where his contracting firm is the only client.

Generally, a separate equipment company is formed and expected to operate on a profitable basis. The cost of equipment, repair, handling, and storage is compared with the income provided by renting to other contractors at a rate comparable to that of other equipment rental companies.

It may be worthwhile to contract with an equipment rental firm to supply all necessary equipment, thus making yard maintenance unnecessary. Hardware and equipment dealers can make a fair profit on supplies and equipment that would cost the contractor more in rehandling and inefficiencies.

8.4 Production Schedules and Progress Reports

Planning prior to construction is essential for completion in the shortest time and at the least cost. Realistic schedules require considerable thought, imagination, and field experience (see Table 11-1).

The production schedule and progress report list the coded items of work (see cost code) to be performed on the project. Estimated production rates and planned sequence of operations are then indicated. Dates of planned operation and actual operation are recorded.

This production schedule can be used to formulate a progress graph illustrating the sequence of operations and when each of the project work items is planned to start, to be underway, and to be completed. It also shows the actual progress of each item of work.

Appendix

Recently, the city of New York adopted a Building Code which may well become the paragon for every new ukase of this type. Its provisions, all developed by consensus among professionals, provide strict specifications for sound construction practice. We include in this volume those excerpts from the New York Building Code which apply to rockwork so that the user may check the methods suggested in this book against applicable standards. The Appendix also includes a section for easy on-site identification of rocks and minerals.

M. D. Morris
Series Editor

PART I

Excerpts from the New York City Building Code*

Article 11 Foundations

Contents

* Data reprinted from Building Code, Local Law No. 76 of the City of New York Effective December 6, 1968, *The City Record*, New York, New York.

Subarticle or Section*	
1102.5	Uplift forces
1102.6	Impact
1102.7	Stability
1103.0	Allowable soil bearing pressures
1103.1	Classification of soil materials
1103.2	Satisfactory bearing material
1103.3	Nominally unsatisfactory bearing material
1103.4	Allowable soil bearing pressures
1103.5	Bearing capacity of nominally unsatisfactory bearing materials
1103.6	Utility services
1104.0	Soil load bearing tests
1104.1	Applicability
1104.2	Procedure
1105.0	Footings, foundation piers, and foundation walls
1105.1	Materials
1105.2	Footings
1105.3	Foundation piers
1105.4	Foundation walls
1105.5	Construction of footings, foundation piers, and foundation walls
1106.0	Pile foundations—general requirements
1106.1	Administrative requirements
1106.2	Minimum pile penetrations
1106.3	Use of existing piles at demolished structures
1106.4	Tolerances and modification of design due to field conditions
1106.5	Minimum spacing of piles
1106.6	Minimum section
1106.7	Capping and bracing of piles
1106.8	Splicing of piles
1106.9	General requirements for installation of piles
1106.10	Use of uncased concrete pile shafts
1106.11	Where more than one pile type, pile capacity, or method of pile installation is used
1106.12	Pile materials
1107.0	Pile foundations—loads
1107.1	Allowable axial load

* "C26" omitted from section numbers in this table.

List of Tables

* "C26" omitted from section numbers in this table.

Sub-Article 1100.0 General

§ C26-1100.1 **Scope.**—The provisions of this article shall establish minimum requirements for the design and construction of the foundations of buildings.

§ C26-1100.2 **Standards.**—The provisions of reference standard RS-11 shall be part of this article.

§ C26-1100.3 **Definitions.**—For definitions to be used in the interpretation of this article, see article 2.

§ C26-1100.4 **Plans.**—For the requirements governing the filing of plans and the work to be shown on the plans, see article 1.

§ C26-1100.5 **Permits.**—For the requirements governing equipment work permits and for equipment use permits, see article 1.

§ C26-1100.6 **General requirements.**—Except as otherwise specifically provided herein, the foundations of buildings including retaining walls and other structures shall bear on, or be carried down to, satisfactory bearing materials in such manner that the entire transmitted load will be distributed over the supporting soils at any depth beneath the foundation at unit intensities within the allowable bearing values established in this article. In addition, foundations shall be proportioned to limit settlements to a magnitude that will not cause damage to the proposed construction or to existing adjacent or nearby buildings during or after construction.

§ C26-1100.7 **Depth of foundations.**—

(a) Footings and Pile Caps.—The bottom surface of any footing, pier, pile cap, or other foundation construction, other than grade beams, shall be carried down at least 4 ft below the lowest level of the adjoining ground or pavement surface that is exposed to frost, except as follows:

(1) In refrigerator, cold storage, or similar areas, the depth shall be increased as required to found the construction below the potential level of freezing in the soil, or loss of heat from the soil shall be prevented by insulation, warm air ducts, circulating systems, or equivalent means.

(2) For foundation elements resting on rock, the footings may rest on the rock surface at shallower depths than those indicated above, provided that visual inspection of the rock surface directly underlying the bearing

area shows it to be free of seams, cracks, or disintegrated materials that could serve as reservoirs for water and thus be subject to freezing.

(3) For foundation elements in the interior of closed and heated buildings, or in cases where the soil underlying the foundation is not subject to frost action, there shall be no mandatory minimum requirement for embedment.

(4) Mobile or portable buildings not more than one story high may be supported on foundation elements bearing at grade, subject to the following conditions:

a. The building shall be supported on jacks, wedges, or other devices that will permit readjustment of level in the event of displacement.

b. The bearing capacity of the underlying soil shall be adequate to support the building loads without rupture. Where the building is to be supported on nominally unsatisfactory bearing materials, the provisions of section 1103.3 and 1103.5 shall apply except that, where the bearing material is confined under pavement and the bearing pressure on the surface of the soil material does not exceed 500 psf, special investigation of the foundation will not be required.

c. The bearing area shall be well drained and not subject to inundation.

d. The levels of the foundations shall be checked and adjusted to compensate for displacements at least once every year, and the owner shall maintain a record of such work, available for inspection by the commissioner.

(5) Where piles project above grade and displacement of the pile cap is prevented by interposing a space between the underside of the cap and the ground, the requirement for embedment of the cap below the level of frost penetration shall not apply.

(b) Grade Beams.—The bottom surface of any grade beam shall be carried down at least 18 in. below the lowest level of the adjoining ground or pavement surface that is exposed to frost.

§ C26-1100.8 **Foundations at different levels.**—Where footings are supported at different levels, or at different levels from the footings of adjacent structures, the influence of the pressures under the higher footings on the stability of the lower footings shall be considered. Consideration shall be given to the requirements for lateral support of the material supporting the higher footings, the additional load imposed on the lower footings, and assessment of the effects of dragdown on adjacent pile-supported buildings.

§ C26-1100.9 **Slabs on grade.**—Slabs on grade within or adjacent to a building shall be so designed to limit settlement of such slabs to a magnitude that will not impair their usability or cause damage to the building or its foundations.

§ C26-1100.10 **Construction.**—The provisions of article 19 relating to safety and of article 10 relating to concrete, timber, masonry, and steel construction shall apply. For inspection requirements, see section C26-1112.0.

(a) Cold Weather.—No foundation shall be placed on frozen soil. No foundation shall be placed in freezing weather unless provision is made to maintain the underlying soil free of frost.

(b) Seepage.—In an excavation where soil and ground water conditions are such that an inward or upward seepage might be produced in soil material intended to provide vertical or lateral support for foundation elements or for adjacent foundations, excavating methods that will control or prevent the inflow of ground water shall be employed to prevent disturbance of the soil material in the excavation or beneath existing buildings. No foundation shall be laid on soil that has been disturbed by seepage unless remedial measures, as directed by an architect or engineer, are taken.

Sub-Article 1101.0 Soil Investigations

§ C26-1101.1 **General.**—Borings in earth or rock, recovery of samples, tests of soil samples, load tests, or other investigations or exploratory procedures shall be performed as necessary for the design and contruction of a safe foundation subject to inspection in accordance with the requirements of sub-article 1112.0.

§ C26-1101.2 **Borings.**—

(a) Number.—At least one boring shall be made for every 2,500 sq. ft. of building area or fraction thereof and, for buildings on piling of such type or capacity that load tests are required, one boring shall be made for every 1,600 sq. ft. of building area or fraction thereof except as indicated in (1) through (3) below.

(1) For one- and two-family dwellings (appurtenant structures such as garages, sheds, and porches shall be considered as part of the dwelling):

a. Buildings supported on footings founded on soil of class 8-65 or better—For contiguous groups of four or more dwellings at least one boring shall be made for every four buildings. For isolated dwellings and for contiguous groups of two or three dwellings, the overall site shall be explored by at least one boring or auger probing or, alternately, the site of each building shall be explored by one test pit carried to a depth of at least 8 ft. below the level of the bottom of the proposed footings.

b. Buildings supported on piles or on footings founded in soil strata of class 9-65, or poorer—The site shall be explored by at least one boring for every four buildings in contiguous groups or one boring for each building in the case of isolated dwellings.

(2) For buildings having a plan area in excess of 10,000 sq. ft. and where sub-surface conditions as determined from preliminary borings or from borings on neighboring sites consist of uniform deposits of materials of class 1-65, 2-65, 3-65, 6-65, 7-65, or 8-65, subject to the approval of the commissioner, the required borings may be reduced to one for every 5,000 sq. ft. of building area or fraction thereof except where the foundation is to be supported on piles.

(3) Where foundations are to rest on rock of class 1-65, 2-65, or 3-65 and such rock is exposed prior to construction over a part or all of the area of the buildings, borings will not be required in those areas where rock is exposed, and the area (within the limits of the building) of the exposed rock surface shall not be included in the area used to compute the required number of borings, provided the following requirements are met:

a. The presence of defects or the inclination of bedding planes in the rock are of such size and location as to not affect the stability of the foundation.

b. The foundation is designed for bearing pressures not exceeding those permitted in table 11-2 without increase for embedment.

(b) Location.—At least two-thirds of the required number of borings shall be located within the area under the building. Those outside the area shall not be more than 25 ft. from the limits of the building. Borings shall be uniformly distributed or distributed in accordance with the loading pattern imposed by the building.

(c) Depth.—

(1) Unless soil material of class 1-65 through 3-65 is encountered at shallower depth, borings shall extend below the deepest part of the excavation as necessary to satisfy the more restrictive of the following requirements:

a. Borings shall extend deep enough into nominally satisfactory bearing material to establish its character and thickness, but not less than the following:

1. Where the soil material is class 5-65—10 ft.
2. For one and two family residences—15 ft.
3. For other cases—25 ft.

b. Borings shall extend to the depth at which the vertical stress caused by the proposed construction is reduced to 10 per cent or less of the original vertical stress at this depth due to the weight of the overburden, except that where strata of soil materials of class 9-65 are encountered within this depth, the borings shall penetrate such strata and be carried to a depth that shows penetration continuous of material of class 8-65 or better as required in a. above.

c. In addition to the requirements of a. and b. above, at least one boring in every 10,000 sq. ft. of building area, but not less than one boring per building, shall be carried to a depth of 100 ft. below the curb for buildings more than one story high and dwellings more than two stories high and for buildings having an average area load exceeding 1,000 psf. The average area load shall be computed as the sum of all dead loads and reduced live loads (sub-article 903.0) of the building, divided by the area of the building at ground level.

(2) Where rock is encountered in borings within the depths noted in (1) the borings shall be cored a minimum of 5 ft. into rock, or farther where necessary or required, to obtain at least a recovery of 35 per cent core from 5 ft. of penetration but not to exceed the depth in (1) above. However, for foundations supported on piles or drilled-in caissons bearing on the rock and having a capacity in excess of 80 tons per unit, the capacity of the rock to support the applied load shall be demonstrated by increasing the length of coring to 10 ft.

(d) Types.—

(1) SOIL BORINGS.—Soil samples shall be recovered at intervals not to exceed 5 ft. and at every change of soil strata. Such samples shall be recovered using a 2 in. O.D. split spoon sampler having an inside diameter of 1⅜ in. and a constant internal cross section. The sampler shall be at least 24 in. long and shall be tipped with a heat treated, sharp cutting shoe. It shall be straight and sound, with an undistorted cross section. The rods to which the sampler is attached shall be 1 in. extra heavy pipe or 1⅝ O.D. drill rods. The sampling tools shall be driven with a 140 lb. hammer having a fall of 30 in. The blows per foot so recorded shall be designated, hereinafter, by the symbol, N. The fall of the hammer shall be a free fall and the energy impact shall not be mitigated by friction of the hoisting line on the drum, friction of the hammer against its guide, or other similar effects. All samples, except those of rock, shall be preserved in air tight bottles having a capacity of at least 8 oz.

(2) ROCK BORINGS.—Where borings are required to penetrate rock they shall be advanced by core drilling, and core samples shall be recovered using a double tube core barrel and diamond bits that provide cores at least 1⅜ in. in diameter.

(e) Data to be Reported.—Records of all borings required by the provisions of section C26-1101.2(a) shall accompany the application for approval of the foundation plans. Such records shall show, as a minimum, the size of casing and the number of blows per ft. required to advance the casing (to the depth that casing is used); the weight of hammer and the distance of fall; a description of the sampler; a description of the drill tools and equipment including, where used, the size of diamond bits and type of core barrels; the number of blows required to drive the sampling spoon for each 6 in. increment of penetration; the elevation of the ground surface referenced to an established datum; the location and depth of the boring and its relation to the proposed construction; the elevations at which samples were taken; the elevations at which core drilling was started and stopped for each "run"; the elevations of the boundaries of soil strata; per cent recovery for each "run" of core drilling; description of the soil strata encountered and geological classification of rock drilled (based on visual examination of cores); any particular, unusual, or special conditions such as loss of water in the earth and rock strata, boulders, cavities, and obstructions, use of special type of samplers, traps, etc.; and the level of ground water together with a description of how and when the ground water level was observed. All abandoned or unsuccessful attempts at borings or rock drilling shall be reported.

(f) Disposition of Samples and Cores.—Soil samples and rock cores shall be retained in an accessible location by the owner, or by the party making application for approval of the foundation plans for a period of one year after the date of issuance of a certificate of occupancy, and shall be available upon reasonable notice for inspection by the commissioner.

§ C26-1101.3 **Test pits.**—Test pits may be substituted for borings on a one-for-one basis. All applicable requirements as to depth, numbers of samples, data to be reported, and disposition of samples shall be observed, except that the use of sample spoon, recording of driving resistance, and recovery of rock cores is not required in test pit investigations.

§ C26-1101.4 **Boring methods.**—Borings shall be made by continuous driving and cleaning out of a pipe casing (including telescoping of smaller sizes inside of larger casing) except as permitted in (a), (b), and (c) following. Where casing is used, it shall be cleaned out to undisturbed soil prior to sampling and the sample spoon driven into soil that has not been affected by chopping, washing, or hydrostatic imbalance.

(a) Uncased Borings.—Uncased borings, including borings where the casing is omitted for part of the depth, may be used if the mud slurry method is followed. The requirements for soil sampling and rock coring shall be the same for uncased borings as for borings made using casing, except that prior to each soil sampling operation the boring shall be substantially cleaned of disturbed material and the sample spoon shall be advanced through any settled solids before counting the blows required to drive the spoon. Longer sample spoons, having a sludge chamber, shall be used where settled solids exceed 6 in. In determining ground water levels, methods shall be used to reduce and replace the mud slurry so that the hydrostatic head may be measured. The procedures shall be described in detail in the records.

(b) Auger Borings.—Borings may be made with augers except that short flight augers shall not be used in granular soils below the water level. Sampling procedures in auger borings for both soil and rock shall be the same as for cased borings. Full hydrostatic head shall be maintained in granular soils below the ground water level during the boring operation.

(c) Maximum Diameter.—Where the bore hole, as drilled by any method, is in excess of 4 in. in diameter, sampling operations shall be performed through a temporary casing having a 4 in. inside diameter or less.

§ C26-1101.5 **Probings and geophysical explorations.—**

(a) FOOTINGS, PIER OR WALL DESIGNS.—Where the foundations for a proposed building consist of footings or foundation piers or walls bearing on rock of class 1-65, 2-65, or 3-65, the use of probings, auger borings or geophysical methods, made without the recovery of soil samples or rock cores, may (except as hereafter specified) be substituted for up to one-half of the number of borings required by the provisions of section C26-1101.2(a), provided that such probings, borings, etc. are carried to adequate depth and are of a nature that will reasonably define the surface contours of the rock. The accuracy of such surface contour definition shall be confirmed by recovering rock cores at the locations of at least one-fifth of the probings or auger borings or, in the case where geophysical methods are used, those borings which are made shall be so distributed as to permit confirmation of the accuracy of the geophysical investigations. The provisions of section C26-1101.2(a) (2), shall not apply where provisions of this section are invoked.

(b) PILE SUPPORT.—Where the foundation for the proposed building consists of piling bearing on rock of class 1-65, 2-65 or 3-65, the provisions of (a) above shall apply, provided that the borings consistently show that the soil overlying the rock consists solely of deposits of class 6-65 through 11-65 and is free of boulders or other obstructions.

(c) GEOPHYSICAL METHODS.—Geophysical investigations shall be conducted by experienced and qualified personnel acceptable to the commissioner who may reject the results and require additional exploration by borings if the results of the geophysical explorations cannot be satisfactorily correlated to the logs of the borings.

§ C26-1101.6 **Existing borings.**—Existing boring data may be utilized subject to the following:

(1) Borings, test pits, probings, etc., that have been made in accordance with all requirements of this section, but not necessarily for the investiga-

tion of the specific project for which application is being made, may be utilized in fulfillment of these provisions.

(2) The logs of borings, test pits, probings, etc., that have been made in accordance with all requirements of this section, but wherein the soil samples and/or rock cores are not available for examination, may be utilized in fulfillment of these provisions to an extent not to exceed one-half of the required number of borings.

(3) Borings, test pits, probings, etc., or the logs thereof, that do not meet the specific requirements of this article, but which are of suitable type and adequate penetration to provide the data required for the safe design and construction of the proposed foundation, may be utilized in fulfillment of the provisions of this section, subject to the approval of the commissioner.

Sub-Article 1102.0 Foundation Loads

§ C26-1102.1 **Soil bearing pressures.**—The loads to be used in computing the bearing pressures on materials directly underlying footings shall be the total column, pier, or wall reactions determined in accordance with the provisions of article 9, on the basis of reduced live load; plus the weight of the foundations; plus the weight of any soil, fill, and slabs on grade that is included within vertical planes projected upward from the extreme limits of the footing to the final ground surface. Live load on grade, or on slabs on grade, within these limits shall also be included. Impact loads shall be considered in accordance with the provisions of section C26-1102.6.

§ C26-1102.2 **Pile reactions.**—The loads to be used in computing pile reactions shall be determined as provided in section 1102.1 except that where piles penetrate fill, clay, silt, peat, or similar compressible strata, the pile loads shall be increased by the amount of drag exerted by such material, and by any overlying strata, during consolidation. Computation of the amount of drag shall consider the amount of added fill, the amount of shear strain between pile (or group) and the soil, the ratio of vertical to horizontal pressure in the soil, and the arrangement of the piles. The soil surrounding or underlying the pile cap shall not be considered as providing any vertical support for the cap.

§ C26-1102.3 **Lateral loads.**—

(a) Earth and ground water pressure.—Every foundation wall or other wall serving as a retaining structure shall be designed to resist, in addition to the vertical loads acting thereon, the incident lateral earth pressures and surcharges, plus hydrostatic pressures corresponding to the maximum probable ground water level.

(b) Wind and other superstructure loads.—Provision shall be made to resist lateral loads imposed on the superstructure due to wind or other causes.

(c) Soil movements.—Buildings shall not be constructed in areas where the soil is subject to lateral movements unless positive provision is made to prevent such movements.

§ C26-1102.4 **Eccentricities.**—Eccentricity of loading in foundations, including eccentricity of loading on the bases of retaining walls, shall be investigated and the maximum soil pressure or pile load (considering eccentricity) shall be kept within the safe capacity thereof as established in sub-articles 1103.0 and 1107.0. Soil pressure and pile load due to eccentricity shall be computed on the basis of straight line distribution of foundation reaction. However, other modes of distribution of the foundation reaction may be assumed, subject to the approval of the commissioner, if it can be demonstrated that the pile and/or soil is capable of sufficient plastic deformation to develop such mode of distribution without failure.

§ C26-1102.5 **Uplift forces.**—Uplift and overturning forces due to wind and hydrostatic pressure shall be considered.

§ C26-1102.6 **Impact.**—Impact forces may be neglected in the design of foundations, except for foundations bearing on loose granular soils, or, regardless of the type of soil material, for foundations supporting cranes, heavy machinery, or moving equipment, or where the ratio of the live load causing impact to the total of the reactions from live load applied without impact plus dead load exceeds one-third.

§C26-1102.7 **Stability.**—The provisions of sub-article 1111.0 shall apply.

Sub-Article 1103.0 Allowable Soil Bearing Pressures

§ C26-1103.1 **Classification of soil materials.**—For purposes of this article, soil materials shall be classified and identified in accordance with table 11-1. In addition, the following supplementary definitions shall apply.

(a) Rock.—

(1) HARD SOUND ROCK.—Includes crystalline rocks such as Fordham gneiss, Ravenswood gneiss, Palisades diabase, Manhattan schist. Characteristics are: the rock rings when struck with pick or bar; does not disintegrate after exposure to air or water; breaks with sharp fresh fracture; cracks are unweathered and less than ⅛ in. wide, generally no closer than 3 ft. apart; core recovery with a double tube, diamond core barrel is generally 85 per cent or greater for each 5 ft. run.

(2) MEDIUM HARD ROCK.—Includes crystalline rocks of (1) above, plus Inwood marble and serpentine. Characteristics are: all those listed in (1) above, except that cracks may be ¼ in. wide and slightly weathered, generally spaced no closer than 2 ft. apart; core recovery with a double tube, diamond core barrel is generally 50 per cent or greater for each 5 ft. run.

(3) INTERMEDIATE ROCK.—Includes rocks of (1) and (2) above, plus cemented shales and sandstone of the Newark formation. Characteristics are: the rock gives dull sound when struck with pick or bar; does not disintegrate after exposure to air or water; broken pieces may show weathered surfaces; may contain fractured and weathered zones up to 1 in. wide spaced as close as 1 ft.; core recovery with a double tube, diamond core barrel is generally 35 per cent or greater for each 5 ft. run.

(4) SOFT ROCK.—Includes rocks of (1) (2) and (3) above in partially weathered condition, plus uncemented shales and sandstones. Characteristics are: rock may soften on exposure to air or water; may contain thoroughly weathered zones up to 3 in. wide but filled with stiff soil; core recovery with a double tube, diamond core barrel is less than 35 per cent for each 5 ft. run, but standard penetration resistance in soil sampling is more than 50 blows per ft. Where core recoveries are less than 20 per cent and the material is to be used for bearing, a minimum 3 in. diameter core shall be recovered and the material recovered shall be classified in accordance with Table 11-1.

(See Table 11-1, a tip-in at pp. 170-171.)

(b) Special Soil Types.—

(1) FINE SAND.—Soils of group SM, containing more than 50 per cent (by weight) of particles passing a no. 60 mesh sieve.

(2) HARDPAN.—Soils of groups GM, GC, and SW, generally directly overlying rock, and which are sufficiently cemented to be difficult to remove by picking.

(3) CLAY SOILS.—Soils of each group SC, CL, and CH shall be classified according to consistency as hard, medium, or soft in accordance with the following:

a. Hard clay.—A clay requiring picking for removal, a fresh sample of which cannot be molded by pressure of the fingers.

b. Medium clay.—A clay that can be removed by spading, a fresh sample of which can be molded by a substantial pressure of the fingers.

c. Soft clay.—A clay, a fresh sample of which can be molded with slight pressure of the fingers.

(4) SILT SOILS.—Soils of each group ML and MH shall be classified as dense, medium, or loose depending on relative difficulties of removal as described for hard, medium, and soft clays in (3) above.

(5) VARVED SILT.—A natural soil deposit consisting of alternating thin layers of silt, clay, and sand in which the silt or silt plus sand layers predominate.

§ C26-1103.2 **Satisfactory bearing material.**—The following materials, or any combination of them shall be considered as generally satisfactory bearing materials; hard sound, medium hard, intermediate, and soft rock; hardpan; granular soils of G and S groups; dense or medium silt soils of groups ML and MH; and hard or medium clay soils of groups CL and CH.

§ C26-1103.3 **Nominally unsatisfactory bearing material.**—Fill material, peat, (group Pt), organic silts and clays (grouping OL and OH), soft or loose soils of groups ML, CL, HM, and CH, varved silt, or satisfactory bearing materials that contain lenses or, or are underlain by, these materials shall be considered as nominally unsatisfactory bearing materials.

§ C26-1103.4 **Allowable soil bearing pressures.**—The allowable bearing pressures on satisfactory bearing materials shall be those established in table 11-2. The allowable bearing pressures on nominally unsatisfactory bearing materials shall be those established in accordance with section C26-1103.5. Allowable bearing pressure shall be considered to be the allowable pressure at a point in the soil mass in excess of the stabilized overburden pressure existing at the same point prior to construction operations. The stabilized overburden pressure existing at a point shall be defined as that portion of the weight of the overlying soil material that is supported by granular interaction rather than pore pressure. In general, the magnitude of the stabilized overburden pressure may be approximated as follows:

(a) The overlying soil material shall have been in place for an adequate length of time to produce a stable condition of pore pressure in, or below, the foundation level. Where the bearing stratum consists of soils of classes 5-65 through 8-65, the bearing stratum shall be considered to be fully consolidated except with regard to the weight of that portion of the overlying soil material that consists of added fill material.

(b) Where all or a portion of the overlying soil consists of fill material, the weight of the fill material shall not be included in the stabilized overburden pressure unless the magnitude of stabilized pressure is verified by an architect or engineer on the basis of laboratory or field tests on undisturbed material.

(c) Where the bearing stratum consists of soils of classes 9-65 through 11-65, the stabilized overburden pressure shall be considered as zero unless the magnitude of the stabilized pressure is established by an architect or engineer on the basis of laboratory or field tests on undisturbed material.

(d) The stabilized overburden pressure shall not include the weight of

Table 11-2 Allowable Soil Bearing Pressures

Class of Material	Description See Note (1)	Basic Allowable Bearing Values (Tons per sq. ft.)—See Notes (10), (11) and (12)	
1-65	Hard Sound Rock	60	See Notes (2) and (8).
2-65	Medium Hard Rock	40	See Notes (2) and (8).
3-65	Intermediate Rock	20	See Notes (2) and (8).
4-65	Soft Rock	8	
5-65	Hardpan	—	See Notes (3) and (8).
6-65	Gravel and Gravel Soils (Soil Groups GW, GP, GM & GC and soils of Soil Groups SW, SP, and SM containing more than 10% of material retained on a No. 4 sieve)	—	See Notes (4) and (8) and (9).
7-65	Sands (other than Fine Sands) (Soil Groups SW, SP & SM but containing not more than 10% of material retained on a No. 4 sieve)	—	See Notes (5), (8) and (9).
8-65	Fine Sand	—	See Notes (6), (8) and (9).
9-65	Clays and Clay Soils (Soil Groups SC, CL & CH)		
	Hard	5	See Note (7).
	Medium	2	See Note (7).
	Soft	See Sec. C26-1103.5.	
10-65	Silts and Silt Soils (Soil Groups ML & MH)		
	Dense	3	
	Medium	1.5	
	Loose	See Sec. C26-1103.5.	
11-65	Nominally Unsatisfactory Bearing Materials	See Sec. C26-1103.5.	

Notes:

(1) Classification.—The soil classifications indicated in this table are those described in section 1103.1. Where there is doubt as to the applicable classification of a soil stratum, the allowable bearing pressure applicable to the lower class of material to which the given stratum might conform shall apply unless the conformance to the higher class of material can be proven by laboratory or field test procedures.

(2) Allowable bearing pressure on rock.—The tabulated values of basic allowable bearing pressures apply only for massive rocks or, for sedimentary or foliated rocks, where the strata are level or nearly so, and, then only if the area has ample lateral support. Tilted strata and their relation to nearby slopes or excavations shall receive special consideration.

(3) Allowable bearing pressure on hardpan.—For hardpan consisting of well cemented material composed of a predominantly granular matrix and free of lenses of fine grained material and inclusions of soft rock, the basic allowable bearing pressure shall be 12 tons per sq. ft. For hardpan consisting of poorly cemented material or

containing lenses of fine grained material, inclusions of soft rock, or a fine grained matrix, the basic allowable bearing pressure shall be 8 tons per sq. ft.

(4) Allowable bearing pressure on gravel and gravel soils.—Values of basic allowable bearing pressure shall be as follows:

(a) For soils of Soil Groups GW, GP, GM, and GC:

Compact, well graded material—10 tons per sq. ft.

Loose, poorly graded material—6 tons per sq. ft.

Intermediate conditions—Estimate by interpolation between indicated extremes.

(b) For soils of Soil Groups SW, SP, and SM, containing more than 10% of material retained on a No. 4 sieve:

Compact, well graded material—8 tons per sq. ft.

Loose, poorly graded material—4 tons per sq. ft.

Intermediate conditions—Estimate by interpolation between indicated extremes.

(5) Allowable bearing pressure on sands.—The basic allowable bearing pressure shall be determined from the resistance to penetration of the standard sampling spoon. The basic allowable bearing pressure in tons per square foot shall equal 0.10 times N but not greater than 6 tons per square foot, nor less than 3 tons per square foot. The appropriate value for the penetration resistance at various areas of the site shall be made by averaging the measured resistance within a depth of soil below the proposed footing level equal to the width of the footing. Where the average values so obtained do not vary by more than 25 per cent of the minimum of the average values over the site of the proposed building, the lowest average value shall be used for the design of the entire building. Where the variation exceeds 25 per cent, the allowable bearing pressure shall be predicated on the lowest average value unless appropriate measures are taken to avoid detrimental amounts of differential settlements of the footings. Where the design bearing pressure on soils of class 7-65 exceeds 3 tons per square foot, the embedment of the loaded area below the adjacent grade shall not be less than 4 feet and the width of the loaded area not less than 3 feet, unless analysis shall demonstrate the proposed construction to have a minimum factor of safety of 20 against shear failure of the soil.

(6) Allowable bearing pressure on fine sand. The basic allowable bearing pressure shall be determined from the resistance to penetration of the standard sampling spoon. The basic allowable bearing pressure in tons per square foot shall equal 0.10 times N but not greater than 4 tons per square foot nor less than 2 tons per square foot, except that, for loose materials (resistance to penetration of the standard sampling spoon 10 blows per foot or less), where the foundation is subjected to vibratory loads from machinery or similar cause, the indicated basic values shall not apply. The allowable bearing pressure shall be established by analysis applying accepted principles of soil mechanics and a report of such analysis satisfactory to the commissioner shall be submitted as a part of the application for the acceptance of the plans.

(7) Allowable bearing pressure on clays and clay soils.—The bearing capacity of medium and hard clays and clay soils shall be established on the basis of the strength of such soils as determined by field or laboratory tests and shall provide a factor of safety against failure of the soil of not less than 20 computed on the basis of a recognized procedure of soils analysis, shall consider probable settlements of the building, and shall not exceed the tabulated maximum values.

(8) Increases in allowable bearing pressure due to embedment of the foundation. —(a) The basic allowable bearing values for rock of classes 1-65, 2-65 and 3-65 shall apply where the loaded area is on the surface of sound rock. Where the loaded area is below the adjacent rock surface and is fully confined by the adjacent rock mass and provided that the rock mass has not been shattered by blasting or otherwise

is or has been rendered unsound, these values may be increased 10 per cent of the base value for each foot of embedment below the surface of the adjacent rock surface in excess of one foot, but shall not exceed twice the basic values. (b) The basic allowable bearing values for soils of classes 5-65 through 8-65 determined in accordance with notes (3), (4), and (5) above, shall apply where the loaded area is embedded 4 ft. or less in the bearing stratum. Where the loaded area is embedded more than 4 ft. below the adjacent surface of the bearing stratum, and is fully confined by the weight of the adjacent soil, these values may be increased 5 per cent of the base value for each foot of additional embedment, but shall not exceed twice the basic values. Increases in allowable bearing pressure due to embedment shall not apply to soils of classes 4-65, 9-65 10-65, or 11-65.

(9) Increase in allowable bearing pressure for limited depth of bearing stratum.—The allowable bearing values for soils of classes 6-65, 7-65 and 8-65 determined in accordance with this table and the notes thereto (including note (8)), may be increased up to one-third where the density of the bearing stratum below the bottom of the footings or the tips of the piles increases with depth provided that: (a) The bearing stratum is not underlain by materials of a lower class. (b) The allowable bearing value of the soil material underlying the bottom of the footings or the tips of the piles increases at least 50 per cent within a depth below the footing or the tips of the piles which is not greater than the width of the footing or the width of the polygon circumscribing the pile group. (c) It shall be demonstrated by a recognized means of analysis that the probable settlement of the foundation due to compression, and/or consolidation do not exceed acceptable limits for the proposed building.

(10) Combination of loads.—The provisions of section 1001.4 of article 10 shall apply.

(11) Correction for foundations bearing on materials of varying bearing capacities.—Where it is show by borings or otherwise that materials of varying bearing value will be used for support of a building: (1) Where the weakest material does not rank below class 5-65, no modifications shall be required. (2) Where the weakest materials rank as classes 6-65 through 8-65, if the difference in basic allowable bearing values for the several materials does not exceed 30 per cent of the allowable value for the poorest material which is to support the foundation, the foundations may be proportioned in direct conformance with the allowable bearing pressures. Where the difference in basic allowable bearing values exceeds 30 per cent, appropriate measures shall be taken to limit the differential settlements of the different portions of the structure to tolerable values. (3) For materials of classes 9-65 and 10-65, in all cases, appropriate measures shall be taken to equalize the settlements of the different portions of the structure.

(12) Inundated footings.—The provisions of this section relating to materials of classes 1-65 through 7-65, shall be deemed equally applicable both to the dry and to the inundated condition of the soil provided: (a) That the subgrade is undisturbed by the construction operations, (b) The bearing area is in a confined condition. For such cases, no reduction in allowable bearing value will be required where the soil supporting the foundation is subject to a rising level of inundation. However, the effects on settlement of a falling level of inundation occurring after construction of the foundation shall be considered in the design.

any soil removed by excavation and not replaced. For footings, the total stabilized overburden pressure shall not exceed the weight of a 1 sq. ft. column of soil (considering submerged weight where the soil column is partly submerged) measured from the bottom of the footing to the lowest level of the final grade above the footing. For a box foundation, where the strength of the slab is adequate to stabilize the underlying soil, the stabilized overburden pressure shall not exceed the weight of a 1 sq. ft. column of soil measured from the bottom of the box to the lowest level of the adjacent grade.

(e) Where the bearing stratum consists of soils of classes 9-65 through 11-65, the allowable bearing pressure shall be adjusted for the effects of rebound due to excavation as determined from consolidation test data.

(f) Where the bearing stratum consists of rocks of classes 1-65 through 3-65, the stabilized overburden pressure shall be neglected.

§ C26-1103.5 **Bearing capacity of nominally unsatisfactory bearing materials.**—Whenever soils exploration shows that the proposed foundation would rest on, or be underlain by, nominally unsatisfactory bearing materials, a report based on soil tests and foundation analysis (including analysis of undisturbed samples) shall be submitted by an architect or engineer, demonstrating, subject to the approval of the commissioner, that the proposed construction, under a condition of 100 per cent overload, is safe against failure of the soil materials. The report shall also show that the probable total magnitude and distribution of settlement to be expected under design conditions will not result in instability of the building or stresses in the structure in excess of the allowable values established in article 10. In addition, the following provisions shall apply:

(a) Fill Materials.—

(1) CONTROLLED FILLS.—Fills shall be considered as satisfactory bearing materials of class 6-65 or 7-65 when placed in accordance with the following procedures, under the provisions for controlled inspection.

a. Before placement of fill, the existing ground surface shall be stripped of all organic growth, timber, rubbish, and debris. After stripping, the ground surface shall be compacted to the density described in d. below.

b. Materials for fill shall consist of sand, gravel, crushed stone, crushed gravel, or a mixture of these, and shall contain no organic matter. The fill materials shall contain no particles exceeding 4 in. in the largest dimension. No more than 30 per cent of the material shall be retained on a ¾ in. sieve. The material passing the ¾ in. sieve shall contain, by weight, no more than 45 per cent passing the 100 sieve, or 12 per cent passing the 200 sieve.

c. The grading of the fill shall be determined in accordance with the applicable procedures of reference standards RS 11-1 and RS 11-2.

d. Fill shall be placed and compacted at its optimum moisture content, in uniform layers not more than 12 in. thick (after compaction) and each layer shall be thoroughly compacted to a density not less than 95 per cent of the density prescribed in reference RS 11-3. The field density shall be verified by in-place density tests made on each lift of the embankment. Fill shall not be placed when frozen or placed on a frozen or wet sub-grade.

(2) UNCONTROLLED FILLS.—Fill material, other than controlled fill, may be used for the support of buildings, other than one- and two-family dwellings, and may be considered as satisfactory bearing material of applicable class, subject to the following:

a. The soil within the building area shall be explored using test pits. At least one test pit, penetrating at least 8 ft. below the level of the bottom of the proposed footings, shall be provided for every 2,500 sq. ft. of building area. Where such test pits consistently indicate that the fill is composed of material that is free of voids and free of extensive inclusions of mud, organic materials such as paper, garbage, cans, or metallic objects, and debris, the provisions of b. and c. below shall apply. Where the test pits show such voids or inclusions, the additional provisions of d. below shall apply. Borings may be used in lieu of test pits, provided that continuous samples ae least 4 in. in diameter are recovered.

b. The building area shall be additionally explored using one standard boring under each column. These borings shall be carried to a depth sufficient to penetrate into natural ground, but not less than 20 ft. below grade, and a sufficient number shall penetrate deeper than 20 ft. as required to meet the provisions of section C26-1101.2. Where such borings show voids or inclusions, the provisions of d. below shall apply.

c. The allowable soil bearing pressure on the fill material shall not exceed 2 tons per sq. ft.

d. Wherever the fill shows void or inclusions as described in a. and b. above either the fill shall be treated as having no presumptive bearing capacity, or the building shall incorporate adequate strength and stiffness to bridge such voids or inclusions or shall be articulated to prevent damage due to differential or localized settlement of the fill.

(b) Organic Silts, Organic Clays, Inorganic Clay, Loose Inorganic Silt, and Varved Silt.—

(1) The allowable bearing pressure shall be determined independently of Table 11-2 subject to the following:

a. For varved silts, the soil bearing pressure produced by the proposed building shall not exceed 2 tons per sq. ft., except that for desiccated or preconsolidated soils, higher bearing pressures will be allowed.

b. For organic silts or clays (groups OL and OH) or for soft or loose soils of groups ML, CL, MH, and CH, the soil bearing pressure produced by the proposed building shall not exceed 1 ton per sq. ft. except that a value of 2 tons per sq. ft. will be permitted on soils that are adequately preconsolidated or artificially treated.

(2) The report required in this section shall contain, as a minimum, the following information:

a. Geological profiles through the area defining the stratigraphy.

b. Sufficient laboratory test data on the compressible material to indicate the coefficient of consolidation, coefficient of compressibility, permeability, secondary compression characteristics, and Atterberg limits.

c. Where the design contemplates improvement of the natural bearing capacity and/or reduction in settlements by virtue of preloading, cross sections showing the amount of fill and surcharge to be placed and design details showing the required time for surcharging shall be indicated, and computations showing the amount of settlement to be expected during surcharging. Records of settlement plate elevations and pore pressure readings, before, during, and after surcharging, shall be filed with the commissioner.

d. The estimated amount and rate of settlement expected to occur after the structure has been completed, including the influence of dead and live loads of the structure.

e. A detailed analysis showing that the anticipated future settlement will not adversely affect the performance of the structure.

f. Where sand drains are to be used, computations showing the diameter, spacing, and method of installation of such drains, shall be provided.

(c) Artificially Treated Soils.—Nominally unsatisfactory soil materials that are artificially compacted, cemented, or preconsolidated (including soils compacted by vibration, cemented by chemical injection, or preconsolidated by use of electric current, but not including cases where preconsolidation consists solely of the use of surcharge with or without sand drains) may be used for the support of buildings, and nominally satisfactory soil materials that are similarly treated may be used to resist soil bearing pressures in excess of those indicated in table 11-2 for the soil in its natural state, subject to the following:

(1) The vertical and lateral extent of the soil that is compacted, cemented, or preconsolidated shall conform to the full extent of the distribution of loading that is assumed for purposes of computing the intensities of the soil bearing pressure. The actual soil bearing pressure shall not exceed the limitations of section C26-1103.5(a) and (b) for nominally satisfactory bearing materials or, for satisfactory bearing materials, shall not exceed the limitations of table 11-2.

(2) After the treatment procedure, a minimum of one boring shall be

made for every 1,600 sq. ft. of that portion of the building area that is supported on treated soil, and a sufficient number of samples shall be recovered from the treated soil to demonstrate the efficacy of the treatment.

§ C26-1103.6 **Utility services.**—Where utility service lines are to be laid in soil materials of class 11-65, provision shall be made to prevent damage to such service lines, as follows:

(a) Where the lines enter a structure, including a building, a manhole, or a junction chamber, that is rigidly supported on piles or in firm bearing material, the services shall be supported on piles or bearing materials of adequate firmness to prevent differential settlement of the service lines with respect to the structure; otherwise, provisions, such as oversized sleeves, flexible connections, utility tunnels, or other approved device, shall be made to permit the anticipated differential movement to occur without damage to the service lines.

(b) Where the lines enter a structure that is supported on soil materials of class 11-65 or on soft clay or loose silt deposits, an engineering analysis shall be made of the probable differential settlement of the utility service line with respect to the structure, and provision shall be made to accommodate such displacement, as described in (a) above.

Sub-Article 1104.0 Soil Load Bearing Tests

§ C26-1104.1 **Applicability.**—Soil load bearing tests made at the option and expense of the owner of the proposed structure and inspected in accordance with the provisions of section C26-106.3, controlled inspection, may be accepted by the commissioner as evidence of allowable bearing capacity of a given soil stratum, subject to the following limitations:

(a) The applicability of soil load bearing tests shall be limited to soil materials of classes 5-65 through 10-65.

(b) Soil load bearing tests shall not be used to justify allowable bearing pressures in excess of the maximum allowable bearing values established in table 11-2 for the applicable class of material.

(c) Soil load bearing tests shall not be applicable where the proposed bearing stratum is underlain by a stratum of lower class, unless analysis indicates that the presence of such lower stratum will not create excessive settlements of the building.

§ C26-1104.2 **Procedure.**—

(a) Preparations.—(1) A sketch showing the layout, levels, number of tests, details of test apparatus, and test procedures shall be filed with the commissioner before conducting such tests. The sketch shall also include the locations and levels of the proposed tests in relation to the contemplated foundation levels as well as the logs of borings that have been made for the building or buildings involved.

(2) The tests shall be made at the levels contemplated for the proposed building footings, and in at least two locations within the limits of the building area. The surfaces at the locations of the proposed tests shall be leveled at the elevations of the proposed test for a clear distance of at least 5 ft. all around the test plate.

(3) The loaded area shall be square and at least 24 in. by 24 in.

(4) When load tests are performed on materials affected by drying, suitable methods shall be used to prevent evaporation from the material.

(5) In the event ground water is present immediately below, at, or above the level required to be tested, dewatering facilities shall be installed to maintain ground water a minimum of 4 ft. below the level of the test plate during the preparation and duration of the test or tests.

(6) Trenches and other provisions at the ground surface shall be made to prevent inflow of surface water and to remove water that may drain into the test area. The entire test area shall be protected against weather and rainfall for the entire duration of the test.

(7) The load plaform providing the support for the test load shall be supported on adequate timber cribbing, which shall not be closer than 5 ft. from the edge of the test plate.

(8) The plate on which the loads are applied shall be of steel having a minimum thickness of 2 in. and shall be set in a depression in the bottom of the pit or loading area about 30 in. square for a 24 in. by 24 in. plate and 8 in. deep. The area on which the plate is placed shall be scraped to remove any disturbed soil. The area to be tested shall be covered with a thin layer of fine clean sand that shall be smoothed and leveled until a completely level surface for placing the test plate is obtained.

(9) Settlement observations shall be made at four corners of test plate by means of dial extensometers. The extensometers shall provide readings to the nearest 0.001 in. In addition, settlement observations shall be taken using an engineer's level reading to 0.001 ft., properly referenced to a well-established benchmark.

(10) All tests shall be made under the surveillance of the commissioner, who shall be duly notified in order that he may be represented.

(b) Loading of the soil.—

(1) Loads shall be applied to the test plate by direct weight or by means of a hydraulic jack. The loading platform or box shall be constructed to provide a concentric load on the plate. If direct weight is employed, the loading increments shall be applied without impact or jar. The weight of the blocking placed on the plate and the loading platform or box

shall be obtained prior to the test and this weight shall be considered as the first increment of load. If a hydraulic jack is employed, facilities for maintaining each increment of desired load constant under increasing settlement shall be provided. The gauge for the jack and the jack shall be calibrated as a unit not more than two weeks prior to the test.

(2) The unit intensity of the test load shall be 1½ times the unit intensity of the loading proposed to be imposed on the soil by the design loads. The test load shall be applied in a minimum of six increments at 25 per cent, 50 per cent, 100 per cent, 125 per cent, and 150 per cent of the proposed working load. Each load increment shall be maintained for a length of time as follows:

a. At the proposed loading and at 150 per cent of load until settlement is less than 0.005 in. over a period of 24 hrs.

b. At other loadings until the settlement is 0.001 in. or less, over a period of 5 min.

(3) Under each load increment, settlement observations shall be made and recorded at intervals of ½ min., 1 min., 4 min., and each 4 min. thereafter after application of load increment except in the instance of the working load and 150 per cent of working load, where, after the 4 min. reading, the time interval shall be doubled successively until the final settlement limitation is reached and the load is increased or removed.

(4) After the test load and limiting rate of settlement under the test load is reached, loads shall be removed in not less than three equal increments and rebound observations made in the same manner as for the loading increments. The final rebound shall be recorded 24 hr. after the entire test load has been removed.

(c) Determination of results.—

(1) Subject to the limitations designated in section 1104.1, the soil load test for soils of classes 5-65 through 8-65 shall be considered as tentatively substantiating the ability of the soil to support the proposed unit intensity of loading if the gross settlement of the test plate under the proposed unit intensity of loading does not exceed ½ in. and the total gross settlement after the 50 per cent overload is applied does not exceed 1 in. This tentative substantiation shall be supported by a report by the architect or engineer on the correlation of the behavior of the load test to the probable behavior of the full size building. The proposed design load shall be considered as acceptable only upon written acceptance by the commissioner.

(2) For soils of classes 9-65 and 10-65, the soil load test alone shall not be considered as evidence of allowable bearing capacity of the soil, but the data so obtained may be utilized to establish an allowable bearing capacity subject to the provisions of section C26-1103.3 and C26-1103.4.

(3) In the event that it is desired to conduct load tests on square areas

larger than the minimum stipulated above, permission may be granted provided notice of such increase in area is properly filed with the test information required to be submitted to the commissioner. The limiting gross settlements stipulated in (1) above shall be increased in relation to the increase in width of test plate in accordance with the following formula:

$$S = \frac{9S_1}{16}\left(\frac{2B}{B+1}\right)$$

where: S = settlement of the larger loaded area (in.)
S_1 = permissible settlement of 24 in. by 24 in. loaded area, as specified in (1) above (in.)
B = side dimension of square plate used for test (ft.)

Sub-Article 1105.0 Footings, Foundation Piers and Foundation Walls

§ C26-1105.1 **Materials.**—All structural elements of foundations as defined in section 1000.6 shall meet the requirements as to type and minimum quality of materials prescribed in article 10. Inspection of materials and construction shall comply with the provisions of section C26-1000.7.

§ C26-1105.2 **Footings.**—Inspections of subgrade of footings, piers and walls shall comply with the requirements of section C26-1112.5.

(a) Wood footings.—Wood footings may be used only for wood frame structures. Wood footings shall be given a preservative treatment in accordance with reference standard RS 11-4.

(b) Pole buildings.—Buildings not more than one story high may be supported on poles embedded in the ground. Wood poles shall conform to the requirements of reference standard RS 11-5 and shall be given a preservative treatment in accordance with reference standard RS 11-4. Steel poles embedded in the soil shall be protected as required by the provisions of section C26-1105.3(c)(1).

(c) Grillages.—Grillage beams shall be provided with proper spacers, stiffeners, and diaphragms, or the space between the beams shall be filled with concrete or grout. In addition, all steel grillages shall be fully protected against corrosion by encasement, coating with metal protection of a type satisfactory to the commissioner or by other approved device.

(d) Design.—

(1) CONCRETE FOOTINGS.—Concrete footings shall be proportioned in accordance with the provisions of reference standard RS 10-3. Reinforcement shall extend to within 4 in. of the edges of the footing.

(2) MASONRY FOOTINGS.—Masonry used for the construction of footings shall be of solid units.

a. Reinforced masonry footings shall meet the requirements of reference standard RS 10-2 and shall be proportioned similarly to the proportioning of reinforced concrete footings.

b. Unreinforced masonry footings shall be of such dimension that a sloping plane extending downward from the top of the footing where it intersects the pier or wall, to the bottom of the footing, and measured at the angle with the horizontal that is indicated below, will be contained entirely within the footing:

Soil bearing capacity of 3 tons per sq. ft. or less—60°

Soil bearing capacity between 3 tons per sq. ft. and 6 tons per sq. ft. —70°

Soil bearing capacity greater than 6 tons per sq. ft.—shall be investigated.

The compressive stress in the footing, based on the assumption that the vertical load is uniformly distributed over horizontal sections bounded by said planes, shall not exceed the values given in reference standard RS 10-1.

§ C26-1105.3 **Foundation piers.**—Foundation piers shall be designed as columns. Reinforced concrete piers shall be designed in accordance with the provisions of reference standard RS 10-3. Reinforced and unreinforced masonry piers shall be designed in accordance with the provisions of reference standards RS 10-2 and RS 10-1. Unreinforced concrete piers shall be designed in accordance with section C26-1105.3(b).

(a) Lateral support.—The equivalent unbraced length of a pier supported by lateral soil pressure may be determined by a recognized method of elastic analysis. Alternatively, such a pier may be assumed to be hinged, but laterally braced at intervals equal to the full height of the pier or eight times the least lateral dimension of the pier, whichever is the lesser value. The provisions of section C26-1105.3(e) shall apply.

(b) Unreinforced concrete piers.—Where unreinforced concrete piers are used, the allowable compressive stress shall not exceed 0.25f′ and the center of cross section of the pier at any level shall not deviate from the line of action of the resultant of all forces (which line of action shall consider the eccentricities due to all loads and moments acting on the pier) by an amount more than 1/60 of its height or 1/10 of its least lateral dimension and the ratio of the height to the least lateral dimension shall not exceed 8. For larger values of the ratio of the height to the least lateral dimensions, or for greater eccentricities, piers shall be reinforced.

(c) Metal shells.—Where piers are encased by a metal shell, the shell may be considered as contributing to the structural strength of the pier provided that the thickness is ⅛ in. or greater, and subject to the following requirements:

(1) Where boring records or site conditions indicate possible deleterious action on the shell, where any portion of the shell is embedded in ash, cinder fill or garbage fill, where the encased piers are used for support of chemical plants, piles of coal, or under other conditions conducive to chemical seepage or corrosive action, or where the encased piers are used for support of electrical generating plants, the metal shells shall be protected against deterioration by encasement, coating, or other device acceptable to the commissioner.

(2) The area of the metal section of the shell multiplied by the efficiency of the horizontal joints shall be considered as equivalent vertical reinforcement of the pier. The area of the metal section of the shell, multiplied by the efficiency of the vertical joints, shall be considered as equivalent spiral reinforcement of the pier.

(d) Minimum dimensions.—The provisions of (e) below shall apply. In addition, the plan dimensions of the pier shall not be less than those of the column above. If the column above is a steel section resting on a base plate, the base plate and pier shall be proportioned for allowable bearing pressures as prescribed in reference standard RS 10-1, RS 10-2, and RS 10-3.

(e) Filling.—The provisions of (a) and (d) above shall apply only where the fill (or backfill) is placed around the pier as controlled fill, and wherein the level of the fill is raised uniformly around the entire perimeter of the pier. Where the fill operation is not controlled the least lateral dimension of a foundation pier shall be 24 in. and the pier shall be proportioned for lateral pressure equal to the pressure of a differential height of fill equal to 5 ft., applied simultaneously with the other loads on the pier specified in sub-article 1102.0.

§ C26-1105.4 **Foundation walls.—**

(a) Concrete.—Concrete foundation walls shall be designed in accordance with the provisions of reference standard RS 10-3 relating to bearing or retaining walls. The equivalent unbraced height of a wall supported by lateral soil pressure may be determined by a recognized method of elastic analysis. Alternatively, such a wall may be assumed to be laterally braced at intervals as follows:

(1) Where fill is placed against both faces of the walls: the full height of the wall or eight times the thickness, whichever is the lesser value.

(2) Where both faces of the wall are not in contact with soil: the height of the unbraced section of the wall or eight times the thickness of the wall, whichever is greater, but not more than the full height of the wall.

(b) Masonry.—Masonry foundation walls shall conform to the provisions of reference standards RS 10-1 and RS 10-2, and the following:

(1) TYPES.—Masonry foundation walls may be of plain or reinforced masonry and shall be of solid units, except that load bearing hollow units will be permitted for support of one- and two-story buildings.

(2) WALL THICKNESS.—Foundation walls of masonry shall be designed and constructed in accordance with the provisions of reference standards RS 10-1 and RS 10-2, supplemented by the provisions of paragraph C26-1105.4(a), relating to unbraced height. The thickness shall be at least 6 in.

§ C26-1105.5 **Construction of footings, foundation piers, and foundation walls.**—The provisions of sections C26-1100.10 and C26-1112.5 shall apply. In addition, methods of installation and construction shall satisfy the following conditions:

(a) Footings, piers, or walls shall be founded on undisturbed soil or on satisfactorily compacted or prepared materials.

(b) Accurate preparation and inspection of the bearing materials directly underlying the foundation shall be possible; and the bearing area shall be substantially level or suitably benched.

(c) Excavation shall be kept substantially free of water during construction of the foundation except that the use of tremie or similar underwater construction will be permitted in the case of foundation bearing directly on rock, provided that the construction procedure will permit thorough cleaning and preparation of the rock surface, and that the surface of the rock is maintained in a clean condition, unfouled by the inflow of soil or settlement of the fluid suspension, until the concrete is in place.

Sub-Article 1106.0 Pile Foundations—General Requirements

§ C26-1106.1 **Administrative requirements.**—

(a) Identification of piles.—A plan showing clearly the designation of all piles by an identifying system shall be filed with the commissioner prior to installation of such piles. All detailed records for individual piles shall bear an identification corresponding to that shown on the plan. A copy of such plan shall be available at the site for inspection at all times.

(b) Record of pile driving.—A record of the penetration and behavior of each pile during installation shall be kept by the architect or engineer designated in section C26-1112.3. Such records shall be prepared on forms furnished by, or satisfactory to, the commissioner and, upon the completion of pile installation, shall be filed with the commissioner together with the records of any additional borings or subsurface information obtained during installation of the piles, and plans showing any deviations of the pile or related constructions (including any corrective measures) from the details and locations shown on the approved plans. Inspection of piling and pile load tests shall conform to the requirements of section C26-1112.3.

§ C26-1106.2 **Minimum pile penetrations.—**

(a) Required by soil bearing capacity.—The provisions of section 1107.1(b)(1) shall apply.

(b) Required for lateral restraint.—The provisions of section C26-1106.7 shall apply.

(c) Piles located near a lot line.—Piles located near a lot line shall be designed on the assumption that the adjacent lot will be excavated to a depth of 10 ft. below the nearest legally established curb level. Where such excavation would reduce the embedded length of the pile, the portion of the pile exposed shall be considered as providing no lateral or vertical support, and the load-carrying determination in accordance with the provisions of sub-article 1107.0 shall be made after the resistance offered by the soil that is subject to potential excavation has been discounted.

§ C26-1106.3 **Use of existing piles at demolished structures.**—Piles left in place where a structure has been demolished shall not be used for the support of new construction unless satisfactory evidence, including load or hammer testing or representative piles, can be produced indicating the capacity, length, and driving conditions of the piles. The load bearing value for such existing piles shall be the least of the values indicated by: (1) the load or hammer test, (2) the capacity of the pile as a structural member, and (3) the allowable bearing pressure on the soil underlying the pile tips, all in accordance with the provisions of sub-article 1107.0.

§ C26-1106.4 **Tolerances and modification of design due to field conditions.—**

(a) Tolerance in alignment of the pile axis.—If the axis of any pile is installed out of plumb or deviates from the specified batter by more than 4 per cent of the pile length, the design of the foundation shall be modified as may be necessary to resist the resulting vertical and lateral forces. In types of piles for which subsurface inspection is not possible, this determination shall be made on the exposed section of the pile, which section, at the time of checking axial alignment, shall not be less than 2 feet. In piles which can be checked for axial alignment below the ground surface, the sweep of the pile axis shall not exceed 4 per cent of the embedded length.

(b) Tolerance in location of the head of the pile.—A tolerance of 3 in. from the designed location shall be permitted in the installation of each pile, without reduction in load capacity of the pile group. Where piles are installed out of position in excess of this amount, the true loading on such piles shall be analytically determined from a survey which defines the actual location of the piles as driven, and using the actual eccentricity in the pile group with respect to the line of action of the applied load. If the total load on any pile, so determined, is in excess of 110 per cent of the allowable load bearing capacity, correction shall be made by installing additional piles or by other methods of load distribution as required to reduce the maximum pile load to 110 per cent of the capacity.

(c) Bent piles.—Where piles have been bent during installation, and the amount of the bend exceeds the allowable tolerance for alignment of the pile axis in (a) above, the condition shall be investigated and, where required, correction made by the installation of additional piles, by strengthening the bent piles, by reduction in capacity or other means acceptable to the commissioner.

§ C26-1106.5 **Minimum spacing of piles.**—Piles shall be spaced to meet the following requirements:

(1) Spacing of piles shall provide for adequate distribution of the load on the pile group to the supporting soil, in accordance with the provisions of section C26-1107.1(b).

(2) In no case shall the minimum center-to-center spacing of piles be less than 24 in., nor less than the values for specific types of piling as indicated in sub-article 1109.0. Unless special measures are taken to assure that piles will penetrate sufficiently to meet the requirements of section C26-1106.2 without interfering with or intersecting each other, the minimum center-to-center spacing of piles shall be twice the average diameter of the butt for round piles; 1.75 times the diagonal for rectangular piles; or, for taper piles, twice the diameter at a level two-thirds of the pile length measured up from the tip. In cases of practical difficulty, the spacing of new piles from existing piles under an adjacent building may be less than the above values provided that the requirements relating to minimum embedment and pile interference are satisfied and that the soil under the proposed and existing buildings is not overloaded by the closer pile grouping.

§ C26-1106.6 **Minimum section.**—Except as provided in sub-article 1109.0 for timber piles, no tapered pile shall be less than 6 in. in diameter at any section, nor have less than an 8 in. diameter at cutoff. The taper of any tapered section may be uniform or may occur in steps. No pile of uniform section shall have a diameter of less than 8 in., or, if not circular, a dimension of less than 7½ in. Tapered shoes or points of lesser dimensions may be attached to the tips of piles.

§ C26-1106.7 **Capping and bracing of piles.**—

(a) Capping of piles.—

(1) EMBEDMENT.—Tops of all piles shall be embedded at least 3 in. in concrete caps. Such concrete shall conform to the provisions of subarticle 1004.0 and shall extend at least 4 in. beyond the edge of all piles. Alternatively, and only where the piles project above the future grade and will be readily accessible for visual inspection at all times, the tops of the piles may be capped with timber or steel caps, which shall be connected to the piles. Cap plates will not be required for steel H piles embedded in a reinforced concrete cap. Inspection of pile caps shall be as required in section C26-1112.4.

(2) UPLIFT.—Where piles are subject to uplift, they shall be anchored into the cap to resist at least 1½ times the amount of such uplift without exceeding the basic allowable stresses as established in article 10.

(3) REINFORCEMENT.—Reinforcement shall be placed to provide at least 3 in. of clear cover, measured to the surface of the pile cap that is in contact with the ground. All reinforcement adjacent to timber or concrete piling shall have a minimum of one inch of concrete protection. Reinforcement shall extend to within 4 in. of the edges of the pile cap.

(4) DESIGN.—Except as modified above, concrete pile caps shall be designed in accordance with the provisions of reference standard RS 10-3.

(b) Bracing of piles.—Except for short piles as described in section C26-1106.7(c), every pile shall be laterally braced to conform with one or more of the following provisions:

(1) RIGID CAP.—Three or more piles connected by a rigid cap shall be considered as being braced provided that the piles are located in radial directions from the centroid of the group not less than 60° apart (within a tolerance of 3 in. in location of the pile.) A two-pile group, in a rigid cap, shall be considered to be braced along the axis connecting the two piles.

(2) BRACE BEAMS.—Piles may be braced by the use of brace beams or ties rigidly connecting to at least two other piles in radial directions not less than 60° nor more than 120° apart. Concrete brace beams shall have minimum dimensions of 1/20 of the clear distance between pile caps, but not less than 8 in. All brace beams shall be proportioned to resist a minimum axial load equal to 3 per cent of the total axial load capacity of the piles that are to be braced by that beam, plus the moment due to any eccentricity between the centroid of the pile group and the line of action of the applied load. Where underlain by soil of class 9-65, 10-65 or 11-65, brace beams shall be proportioned to support the weight of soil, slab-on-ground, and live load on the slab-on-ground that is contained within vertical planes projected upward from the lateral limits of the brace beam. The design of brace beams to resist these loads shall conform to the provision of article 10.

(3) CONCRETE SLAB-ON-GRADE.—A continuous concrete slab or mat on grade, that is 5 in. or more in thickness and reinforced, and that extends at least 40 ft. in each direction and is anchored to the pile caps (or in which the piles are embedded at least 3 in.), may be used in lieu of brace beams for bracing of pile caps, providing that the slab is supported on material having an allowable bearing pressure of 1½ tons per square foot or better and such material is not underlain by nominally unsatisfactory bearing materials.

(4) OTHER MEANS.—Piles may be braced by anchors, anchor wall, or other means acceptable to the commissioner.

(5) FLOOR SYSTEMS.—Single-pile or two-pile groups or a single line of piles may be considered to be adequately braced if connected to, and braced by, a self supporting floor system provided: (1) that the details and dimensions of the floor and the wall or pier are of adequate strength to resist lateral displacement of the pile cap under conditions of maximum eccentricity of the applied load; and (2) that the wall or pier is braced until connection of the floor framing is made and the flooring (or slab) is in place.

(6) SPECIAL REQUIREMENTS FOR BRACING BATTER PILES.—The provisions of paragraphs (1) to (5) above shall apply. In addition, provisions shall be made to oppose the lateral thrust resulting from the pile inclination.

(c) Bracing of short piles.—

(1) All pile caps supported by piles that penetrate less than 10 ft. below cutoff level or less than 10 ft. below ground level shall be braced against lateral movement. Such bracing may consist of connection to other piles embedded more than 10 ft. below those levels; the use of suitable anchors, connection to a slab-on-grade or the floor system as described in (b)(3) and (b)(5) above, or by other equivalent means. The heads of the piles shall be fixed in the cap.

(2) Where the embedded length of piles located near a lot line would be reduced to less than 10 ft. by excavation of the adjacent site to a depth of 10 ft. below the nearest established curb level, the provisions of (1) above shall apply.

§ C26-1106.8 **Splicing of piles.**—Splices shall be constructed so as to provide and maintain true alignment and position of the component parts of the pile during installation and subsequent thereto, and shall be of adequate strength to transmit the vertical and lateral loads (including tensions) and the moments occurring in the pile section at the location of the splice without exceeding the allowable stresses for such materials as established in section C26-1107.1(a). Except for piles which can be visually inspected after driving splices shall develop at least 50 per cent of the capacity of the pile in bending. In addition, all pile splices in the upper 10 ft. of the pile section shall be capable of resisting (at allowable working stresses) the moment and shear that would result from an assumed eccentricity of the pile load of 3 in., or the pile shall be braced in accordance with the provisions of section C26-1106.7 to other piles that do not have splices in the upper 10 ft. of embedment. For piles located near a lot line, the embedded length of such piles shall be determined on the basis that the adjacent site will be excavated to a depth of 10 ft. below the nearest established curb level.

§ C26-1106.9 **General requirements for installation of piles.—**

(a) Protection of adjacent property.—Piles shall be installed with adequate provision for the protection of adjacent buildings and property.

(b) Protection of the pile during installation.—Piling shall be handled and installed to the required penetration and resistance by methods that leave their strength unimpaired and that develop and retain the required load-bearing resistance. Any damaged pile shall be satisfactorily repaired or the pile shall be rejected. Subject to the approval of the commissioner, damaged piles may be used at a fraction of the design load as determined by the architect or engineer in lieu of repair or rejection.

(c) Protection of pile materials after installation.—Where boring records or site conditions indicate possible deleterious action on pile materials due to soil constituents, changing water levels, or other causes, such materials shall be adequately protected by preservatives or encasements that will not be rendered ineffective by driving and that will prevent such deleterious action. The following specific provisions shall apply:

(1) Untreated timber piles shall not be used unless the top level of the pile is below the permanent water table. The permanent water table level shall not be assumed higher than the invert level of any sewer, drain, or subsurface structure in the adjacent streets, nor higher than the water level at the site resulting from the lowest drawdown of wells or sumps, but in no case shall untreated timber piles be used where the cut-off level is less than 10 feet below the adjacent legal grade. Where treated piles are required, preservative treatment shall consist of impregnation with creosote solution or, for piles entirely embedded below grade, a pentachlorophenal solution may be used. Treatment shall be in accordance with all requirements of reference standard RS 11-6.

(2) Piles installed in ash or garbage fills, cinder fills, or which are free-standing in or near a seawater environment, or which are used for the support of chemical plants, coal piles or under similar conditions of chemical seepage or aggressive action, or which are used for support of electrical generating plants, shall be investigated regarding the need for special protective treatment and, where protective treatment is indicated, shall be protected against deterioration by encasement, coating, or other device acceptable to the commissioner.

(d) Equipment.—Equipment and methods for installing piles shall be such that piles are installed in their proper position and alignment, without damage. Equipment shall be maintained in good repair.

§ C26-1106.10 **Use of uncased concrete pile shafts.**—The use of uncased shafts (i.e., where the concrete of the pile shaft is in direct contact with the surrounding soil) will be permitted under the following conditions:

(a) For bored piles.—Where the bored hole is maintained free of water before and during placement of the concrete and the sides and bottom can be inspected prior to such placement, and provided that no displacement pile shall be installed within 15 ft. of any bored pile.

(b) For driven piles (including all piles wherein installation utilizes a temporary casing).—The maximum length of the uncased shaft shall be limited to 10 ft. unless otherwise permitted by the commissioner. Uncased shafts will be permitted to be formed in soil below the water table but shall not be formed in any soil of class 9-65 that is of medium or soft consistency; in any soil of class 10-65 that is of medium or loose density; or in any soil of class 11-65.

§ C26-1106.11 **Where more than one pile type, pile capacity, or method of pile installation is used.**—Wherever it is proposed to: (1) construct a foundation for a building utilizing piles of more than one type or capacity; (2) modify an existing foundation by the addition of piles of a type or capacity other than those of the existing piling; (3) construct or modify a foundation utilizing different methods or more than one method of installation, or using different types or capacities of equipment (such as different types of hammers having markedly different striking energies or speeds); or (4) support part of a building on piles and part on footings, the several parts of the building supported on the different types, capacities, or modes of piling shall be separated by suitable joints providing for differential movement, or a report shall be submitted by the architect or engineer establishing to the satisfaction of the commissioner that the proposed construction is adequate and safe, and showing that the probable settlements and differential settlements to be expected will not result in instability of the building or stresses in the structure in excess of the allowable values established in article 10. The provisions of section C26-1107.1 (d) relating to required load tests shall apply separately and distinctly to each different type or capacity of piling, method of installation, or type or capacity or equipment used, except where analysis of the probable, comparative behavior of the different types or capacities of the piles or the methods of installation indicates that data on one type or capacity of pile permits a reliable extrapolation of the probable behavior of the piles of other types and capacities.

§ C26-1106.12 **Pile materials.**—The provisions of section C26-1000.1

and C26-1000.9 relating to "classification of materials, assemblies and methods of construction" and to the use of "used and unidentified materials" shall apply.

Sub-Article 1107.0 Pile Foundations—Loads

§ C26-1107.1 **Allowable axial load.**—The allowable axial load on a pile shall be the least value permitted by consideration of the following factors (for battered piles, the axial load shall be computed from the resultant of all vertical loads and lateral forces occurring simultaneously):

1. The capacity of the pile as a structural member.
2. The allowable bearing pressure on soil strata, underlying the pile tips.
3. The resistance to penetration of the piles, including resistance to driving, resistance to jacking, the rate of penetration, or other, equivalent criteria as established in this section.
4. The capacity as indicated by load test, where load tests are required.
5. The maximum loads prescribed in section C26-1107.1(e).

(a) The capacity of the pile as a structural member.—

(1) EMBEDDED PORTION OF THE PILE.—The compressive stress on any cross section of a pile produced by that portion of the design load that is considered to be transmitted to that section shall not exceed the allowable values for the construction materials as established in table 11-3. The tensile stress shall not exceed the values established in article 10 for like material.

(2) PORTION OF THE PILE THAT IS NOT EMBEDDED.—That portion of any pile that is free standing in air or water shall be designed as a column in accordance with the provisions of article 10, fixed at a point 5 ft. below the soil contact level in class 8-65 material or better and 10 ft. below in any other material. The conditions of lateral and rotational restraint offered by the pile cap shall be considered in determining the equivalent unbraced length.

(3) LOAD DISTRIBUTION ALONG EMBEDDED PORTION OF THE PILE.—The portion of the design load acting on any cross-section of a pile may be determined by analysis, considering time dependent changes in distribution of the load. As an alternative method for the purposes of this section, it may be assumed that:

a. For piles embedded 40 ft. or more in materials of class 10-65 or better, or in controlled fills, and bearing on or in materials of classes 1-65 to 5-65: 75 per cent of the load shall be assumed to be carried by the tip. For shorter piles, with similar conditions of embedment and bearing, 100 per cent of the load shall be assumed carried by the tip.

b. For piles embedded in materials of class 10-65 or better, or in controlled fills, and bearing on or in materials of classes 6-65 to 10-65 (or controlled fills): the full load shall be assumed to act at a cross section located at ⅔ of the embedded length of the pile measured up from the tip. Where tapered piles are used, the stress at all sections of the pile shall be determined on the basis that the full load acts at a location as dscribed above and that ⅓ of the full load acts at the tip. The stresses so computed shall not exceed the allowable values in table 11-3.

c. For conditions not covered in a. and b. above the provisions relating to analysis shall apply.

(b) Allowable bearing pressure on soil strata underlying the pile tips.—

(1) BEARING CAPACITY.—The allowable pile load shall be limited by the provision that the pressures in materials at and below the pile tips produced by the loads on individual piles and by the aggregate of all piles

Table 11-3 Allowable Compressive Stress for Pile Materials

Pile Material	Allowable compressive stress
Concrete	Concrete—The provisions of Reference Standard RS 10-3 relating to short compression members shall apply. For working stress design use $0.25f'_e$. For ultimate strength design use minimum eccentricity of 5 per cent, $\phi = 0.70$ ($\phi = 0.75$ where a permanent metal shell having a minimum wall thickness of ⅛ inch is used), and load factors as specified in Reference Standard RS 10-3. The above provisions shall be deemed applicable to reinforced and unreinforced section. For unreinforced sections use $D_8 = D$ and $d = t$ (symbols refer to those used in Reference Standard RS 10-3).
	Reinforcing steel—$0.40f_y$ but not greater than 30,000 psi. (The provisions of sub-article 1004.0 relating to ties, spirals, and percentages of reinforcing steel for reinforced concrete compression members shall apply.)
Timber	See timber piles (section C26-1109.2).
Steel	H piles—$0.35f_y$, with f_y not to be taken as greater than 36,000 psi. Minimum thickness of metal shall be 0.40.
	Pipe piles, shells for cast-in-place concrete piles and shells of pipe sections used in caisson piles: $0.35f_y$, (f_y not to be taken as greater than 36,000 psi.) for thicknesses of ⅛ in. or more. Metal thinner than ⅛ in. shall not be considered as contributing to the structural strength of the pile section.
Core sections for caisson piles: $0.50f_y$, with f_y not to be taken as greater than 36,000 psi.	

NOTE:

f'_e = 28 day compressive strength of concrete.

f_y = Minimum specified yield strength of steel.

in a group or foundation, shall not exceed the allowable bearing values established in sub-article 1103.0. The provisions of section C26-1103.4 and C26-1103.5 shall apply. The transfer of load from piles to soil shall be determined by a recognized method of analysis. As an alternative, for purposes of this section, pile or pile groups may be assumed to transfer their loads to the underlying materials by spreading the load uniformly at an angle of 60° with the horizontal, starting at a polygon circumscribing the piles, located as follows:

a. For piles embedded entirely in materials of classes 4-65 to 8-65, or in controlled fill materials, the polygon shall be circumscribed at a level located ⅔ of the embedded length of the pile, measured up from the tip.

b. For piles penetrating through soils of classes 9-65, 10-65, or 11-65 into bearing in soils of class 8-65 or better, the polygon shall be circumscribed at the bottom of the strata of class 9-65, 10-65 or 11-65 materials.

c. In the case of piles having enlarged bases, the lateral distribution of the load to the soil may be assumed to begin at the junction of the shaft and the enlarged base and to extend as follows:

1. In the case where the enlarged base is formed in loose or medium compact (N value less than 30) soils of class 6-65 or 7-65 that extend 20 ft. or more below the junction of the base and shaft, or that are of lesser extent but are directly underlain by soil of class 5-65 or better, the bearing area may be taken at a plane 6 ft. below said junction but not lower than the bottom of the soil strata of class 6-65 or 7-65.

2. Where the enlarged base is formed in compact (N value 30 to 60) soils of class 6-65 or 7-65, or in any soil of these classes that extends less than 20 ft. below the junction of the base and the shaft and that is underlain by soil of class 8-65 or poorer, the bearing area shall be taken at planes less than 6 ft. below said junction, with a lower limit of 3 ft. where the material is very compact (N value 60, or greater) and the extent of the class 6-65 or 7-65 material is 10 ft. below the junction of shaft and base. (The provisions of section C26-1109.5(e) relating to minimum depth of bearing stratum below the junction of base and shaft shall apply.) For conditions intermediate between that described in 1. above and the lower limit conditions described here, the location of the bearing area may be determined by linear interpolation between the indicated limits of N value and extent of bearing material below the junction of shaft and base, giving equal weight to both variables.

3. Where the enlarged base is formed in or on soils of class 4-65 or 5-65, the bearing area shall be taken at a depth below the junction

of the shaft and base consonant with the size and depth of the base formed, and as evaluated from the required test piles.

d. For all piles bearing on soils of classes 1-65 to 3-65, analysis of load distribution will not be required if the requirements relating to capacity of the pile as a structural member, to resistance to penetration, to load test where required, and to maximum tabulated loads are satisfied.

e. For piles bearing in soils of classes 9-65 and 10-65, for cases not described above, or for any case where the method of installing the pile utilizes a temporary casing, the provision relating to analysis shall apply.

f. In no case shall the area considered as supporting the load extend beyond the intersection of the 60° planes of adjacent piles or pile groups.

(2) BEARING STRATUM.—The plans for the proposed work shall establish, in accordance with the requirements relating to allowable bearing pressure, the bearing strata to which the piles in the various sections of the building are to be penetrated and the approximate elevations of the top of such bearing strata. Where penetration of a given distance into the bearing strata is required for adequate distribution of the loads, such penetration shall be shown on the plans. The indicated elevations of the top of the bearing strata shall be modified by such additional data as may be obtained during construction. All piles shall penetrate to or into the designated bearing strata.

(c) Capacity as indicated by resistance to penetration.—Where subsurface investigation, as described in sub-article 1101, or general experience in the area, indicates that the soil that must be penetrated by the pile consists of glacial deposits containing boulders, or fills containing riprap, excavated detritus, masonry, concrete, or other obstructions in sufficient numbers to present a hazard to the installation of the piles, the selection of type of pile and penetration criteria shall be subject to the approval of the commissioner but in no case shall the minimum penetration resistance be less than that stipulated in Tables 11-4 and 11-5.

(1) PILES INSTALLED BY USE OF STEAM-POWERED, AIR-POWERED, DIESEL-POWERED OR HYDRAULIC IMPACT HAMMERS.—

a. The minimum required driving resistance and the requirements for hammer energies for various types and capacities of piles are given in Tables 11-4 and 11-5. To obtain the required total driving resistance, the indicated driving resistance shall be added to any driving resistance experienced by the pile during installation, but which will be dissipated with time (resistance exerted by non-bearing materials or by materials which are to be excavated). For purposes of this section, the resistance exerted by non-bearing materials may be approximated as the resistance

Table 11-4 Minimum Driving Resistance and Minimum Hammer Energy for Steel H-Piles, Pipe Piles, Precast and Cast-in-Place Concrete Piles and Composite Piles (other than timber)

		Minimum Driving Resistance[a, c, e]				
Pile Capacity (tons)	Hammer[b] Energy (ft. lbs.)	Friction Piles (blows/ ft.)	Piles Bearing on Hardpan (Soil Class 5-65) (blows/ ft.)	Non-Displacement Piles Bearing on Decomposed Rock (Soil Class 4-65) (blows/ ft.)	Displacement Piles Bearing on Decomposed Rock (Soil Class 4-65) (blows/ ft.)	Piles Bearing on Rock (Soil Classes 1-65, 2-65, & 3-65)
Up to 20	15,000	19	19	48	48	
	19,000	15	15	27	27	
	24,000	11	11	16	16	
30	15,000	30	30	72	72	
	19,000	23	23	40	40	
	24,000	18	18	26	26	
40	15,000	44	50	96	96	
	19,000	32	36	53	53	
	24,000	24	30	34	34	
50	15,000	72	96	120	120	5 Blows per ¼ inch (Minimum hammer energy of 15,000 ft. lbs.)
	19,000	49	54	80	80	
	24,000	35	37	60	60	
	32,000	24	25	40	40	
60	15,000	96		240	240	
	19,000	63		150	150	
	24,000	44		100	100	
	32,000	30		50	50	
70 & 80	19,000					
	24,000					
	32,000					
			5 Blows per ¼ inch (Minimum hammer energy of 15,000 ft. lbs.)	5 Blows per ¼ inch (Minimum hammer energy of 19,000 ft. lbs.)		
100						
Over 100						

NOTES:

[a] Final driving resistance shall be the sum of tabulated values plus resistance exerted by non-bearing materials. The driving resistance of non-bearing materials shall be taken as the resistance experienced by the pile during driving, but which will be dissipated with time and may be approximated as described in Section C26-1107.1(c)(1)a.

[b] The hammer energy indicated is the rated energy.

[c] Sustained driving resistance—where piles are to bear in soil classes 4-65 and 5-65, the minimum driving resistance shall be maintained for the last 6 inches, unless a higher sustained driving resistance requirement is established by load test. Where piles are to bear in soil classes 6-65 through 10-65, the minimum driving resistance shall be maintained for the last 12 inches unless load testing demonstrates a requirement for higher sustained driving resistance. No pile need be driven to a resistance to penetration (in blows per inch) more than twice the resistance indicated in this table, nor beyond the point at which there is no measurable net penetration under the hammer blow.

[d] The tabulated values assume that the ratio of total weight of pile to weight of striking part of hammer does not exceed 3.5. If a larger ratio is to be used, or for other conditions for which no values are tabulated, the driving resistance shall be approved by the commissioner.

[e] For intermediate values of pile capacity, minimum requirements for driving resistance may be determined by straight line interpolation.

Table 11-5 Minimum Driving Resistance and Hammer Energy for Timber Piles

Pile Capacity (tons)	Minimum Driving Resistance (blows-in.) to be added to driving resistance exerted by non-bearing materials[a, c, d]	Hammer Energy (ft.-lbs.)[b]
Up to 20	Formula in Note [d] shall apply	7,500-12,000
Over 20 to 25		9,000-12,000
		14,000-16,000
Over 25 to 30		12,000-16,000 (single-acting hammers)
		15,000-20,000 (double-acting hammers)
Greater than 30*		

NOTES:

[a] The driving resistance exerted by non-bearing materials is the resistance experienced by the pile during driving, but which will be dissipated with time and may be approximated as described in Section C26-1107.1(c)(1)a.

[b] The hammer energy indicated is the rated energy.

[c] Sustained driving resistance.—Where piles are to bear in Soil Classes 4-65 and 5-65, the minimum driving resistance shall be maintained for the last six inches, unless a higher sustained driving resistance requirement is established by load test. Where piles are to bear in Soil Classes 6-65 through 10-65, the minimum driving resistance measured in blows per inch shall be maintained for the last 12 inches unless load testing demonstrates a requirement for higher sustained driving resistance. No pile need be driven to a resistance to penetration (in blows per inch) more than twice the resistance indicated in this Table nor beyond the point at which there is no measurable net penetration under the hammer blow.

[d] The minimum driving resistance shall be determined by the following formula:

$$P = \frac{2W_hH}{s + 0.1} \text{ or } P = \frac{2E}{s + 0.1}$$

where:

P = Allowable pile load in pounds.
W_p = Weight driven in pounds.
W_h = Weight of striking part of hammer in pounds.
H = Actual height of fall of striking part of hammer in feet.
E = Rated energy delivered by the hammer per blow in foot lbs.
s = Penetration of pile per blow, in inches, after the pile has been driven to a depth where successive blows produce approximately equal net penetration.

The value $\frac{W_p}{W_h}$ shall not exceed 3.

* Appears with no corresponding hammer energy in the original text, apparently should be deleted.

to penetration of the pile recorded when the pile has penetrated to the bottom of the lowest stratum of nominally unsatisfactory bearing material (class 11-65, but not controlled fill) or to the bottom of the lowest stratum of soft or loose deposits of class 9-65 or 10-65 but only where such strata are completely penetrated by the pile. The provisions of sub-articles 1108.0 and 1109.0 shall also apply.

b. Alternate for similitude method.—The requirement for installation of piling to the penetration resistances given in tables 11-4 and 11-5 will be waived where the following five conditions prevail:

1. The piles bear on, or in, soil of class 5-65 through class 10-65.

2. The stratigraphy, as defined by not less than one boring for every 1600 sq. ft. of building area, shall be reasonably uniform or divisible into areas of uniform conditions.

3. Regardless of pile type or capacity, one load test, as described in section C26-1107.1(d), shall be conducted in each area of uniform conditions, but not less than two typical piles for the entire foundation installation of the building or group of buildings on the site, nor less than one pile for every 15,000 sq. ft. of pile foundation area shall be load tested.

4. Except as permitted by the provisions of 6 below, all building piles within the area of influence of a given load-tested pile of satisfactory performance shall be installed to the same or greater driving resistance as the successful load-tested pile. The same or heavier equipment of the same type that was used to install the load-tested pile shall be used to install all other building piles, and the equipment shall be operated identically. Also, all other piles shall be of the same type, shape, external dimension, and equal or greater cross-section as the load-tested pile. All building piles within the area of influence represented by a given satisfactory load-tested pile shall bear in, or on the same bearing stratum as the load test pile.

5. A report by an architect or engineer shall be submitted establishing to the satisfaction of the commissioner, that the soil bearing pressures do not exceed the values permitted by the provisions of sub-article 1103.0 and that the probable differential settlements will not cause stress conditions in the building in excess of those permitted by the provisions of article 10.

6. Where the structure of the building and length of the piling is such as to cause the building and its foundation to act as an essentially rigid body, the building piles may be driven to length and/or penetration into the bearing stratum without regard to penetration resistance, subject to the requirement of paragraph 5, relating to submission of report.

(2) PILES INSTALLED BY JACKING OR OTHER STATIC FORCES.—The carrying capacity of a pile installed by jacking or other static forces shall be not more than 50 per cent of the load or force used to install the pile to the required penetration, except for piles jacked into position for underpinning. The working load of a temporary underpinning pile shall not exceed the total jacking force at final penetration. The working load of each permanent underpinning pile shall not exceed the larger of the following values: ⅔ of the total jacking force used to obtain the required penetration if the load is held constant for 7 hr. without measurable settlement; or ½ of the total jacking force at final penetration if the load is held for a period of 1 hr. without measurable settlement. The jacking resistance used to determine the working load shall not include the resistance offered by non-bearing materials which will be dissipated with time.

(3) PILES INSTALLED BY USE OF VIBRATORY HAMMER.—The capacity of piles installed by vibratory hammer shall not exceed the value established on the principle of similitude, as follows:

a. Comparison piles, as required by the provisions of section C26-1107.1(d), shall be installed using an impact hammer and driving resistances corresponding to the proposed pile capacities as determined in paragraph C26-1107.1(c)(1) or to tip elevations and driving resistances as determined by the architect or engineer.

b. For each comparison pile, install an identical index pile by use of the vibratory hammer at a location at least 4 ft., but not more than 6 ft., from each comparison pile. The index piles shall be installed to the same tip elevation as the comparison pile, except that where the comparison piles bear on soils of classes 1-65 to 5-65, the index piles shall bear in, or on, similar material. All driving data for the index piles shall be recorded.

c. The index piles shall be load tested in accordance with the provisions of section C26-1107.1(d). Should the specified load test criteria indicate inadequate capacity of the index piles, steps a, b, and c shall be repeated using longer, larger, or other types of piles.

d. All building piles within the area of influence of a given, satisfactorily tested index pile shall be installed to the same or lesser rate of penetration (in. per min.) as of the successful index pile. The same equipment that was used to install the index pile, identically operated as to rpm, manifold pressure, etc., shall be used to install the building piles. Also, all building piles shall be of the same type, size, and shape as the index pile. All building piles within the area of influence as represented by a given satisfactorily tested index pile shall bear in, or on, the same bearing stratum as the index pile.

Table 11-6 Basic Maximum Pile Loads

Type of pile	Basic maximum pile load (tons)
Caisson piles	No upper limit
Open-end pipe (or tube) piles bearing on rock of classes 1-65, 2-65, and 3-65	18 in. O.D. and greater—250 less than 18 in. O.D.—200
Closed-end pipe (or tube) piles, H piles, cast-in-place concrete and compacted concrete piles bearing on rock of classs 1-65, 2-65 and 3-65	150
Piles (other than timber piles) bearing on soft rock (class 4-65)	
1) Displacement piles such as pipe, cast-in-place concrete, and compacted concrete piles	60
2) Non-displacement piles such as open-end pipe and H piles	80
Piles (other than timber piles) bearing on hardpan (class 5-65) overlying rock	100
Piles (other than timber piles) that receive their principal support other than by direct bearing on soils of classes 1-65 to 5-65	60
Timber piles	
Bearing in soils of classes 1-65 to 5-65	25
Bearing in soils of classes 6-65 to 10-65	30

(d) Capacity as indicated by load test.—Load test of piling shall be required as follows:

(1) PILES INSTALLED BY STATIC FORCES.—The load bearing capacity of all types and capacities of piles installed by static forces, other than caisson piles and underpinning piles, shall be demonstrated by load test.

(2) PILES DRIVEN BY IMPACT HAMMERS.—The load bearing capacity of piles installed by impact hammers shall be demonstrated by load test when the proposed pile capacity exceeds the following values:

a. Caisson piles—No load test required.

b. Piles installed open end to rock of class 1-65, 2-65 or 3-65—100 tons, except as provided in paragraph d, and except that no load tests will be required for piles up to 200 tons capacity wherein the pile load does not exceed 80 per cent of the load determined on the basis of limiting stresses in the pile materials and provided that the pipe or shell be driven to the resistance indicated in Table 11-4.

c. Piles bearing on rock or hardpan (soil classes 1-65 to 5-65) other than as described in paragraph b above, and except as provided in paragraph d below—40 tons.

d. Piles bearing on materials of class 8 or better, wherein, on the assumption that 100 per cent of the load reaches the pile tip (or, in the case of piles having an enlarged base or other enlargement of the bearing area, the top of the enlargement) the bearing pressure on the soil underlying the tips or bases can be demonstrated to be equal to or less than the values of basic allowable pressure indicated in Table 11-2—provided that the class and density of the bearing material supporting the piles be confirmed by not less than one boring at each column location, then the commissioner may reduce the required number of load tests.

e. All other types of piles—30 tons.

(3) PILES INSTALLED BY USE OF VIBRATORY HAMMERS.—The load bearing capacity of all types and capacities of piles (other than caisson piles) shall be demonstrated by load test.

(4) LOAD TEST PROCEDURES.—Before any load test is made, the proposed apparatus and structure to be used in making the load test shall be satisfactory to the commissioner and when required by him, all load tests shall be made under his surveillance or that of his representative. A complete record of such tests shall be filed with the commissioner.

a. Areas of the foundation site within which the subsurface soil conditions are substantially similar in character shall be established. In addition, for friction piles bearing in, or on, soil materials of classes 6-65, or poorer, the uniformity of each such area shall be verified by installing at least three penetration-test piles, distributed over the area. Continuous records of penetration resistance shall be made for such piles. If the records of penetration resistance are not similar or are not in reasonable agreement with the information obtained from the borings, the assumed areas of similar subsurface conditions shall be modified in accordance with the information derived from the penetration-test piles and additional penetration-test piles shall be installed as required to verify the uniformity of such areas.

b. For piles installed by jacking or other static forces or by impact hammer, one load test shall be conducted in each area of uniform conditions, but not less than two typical piles for the entire foundation installation of the building or group of buildings on the site, and not less than one pile for each 15,000 sq. ft. of the area of the building wherein said piles are to be used shall be load tested. For piles installed by use of vibratory hammers, one comparison pile shall be installed and one index pile shall be load tested in each area of uniform conditions, but not less than two index piles shall be tested for the entire foundation installation of the building or group of buildings on the site, nor less than one index pile be tested for every 7,500 sq. ft. of pile foundation area. For piles whose capacity is determined on the basis of similitude, the provisions of section C26-1107.1(c)(1)b. shall apply.

c. The load test shall be conducted by a method that will maintain constant load under increasing settlement. Settlement observations shall be made by means of dial extensometers. The extensometers shall provide readings to the nearest 0.001 in. In addition, settlement observations shall be taken using an engineer's level reading to 0.001 ft., properly referenced to a well-established benchmark.

1. Test loads shall be applied by direct weight or by means of a hydraulic jack. The loading platform or box shall be carefully constructed to provide a concentric load on the pile. If direct weight is employed, the loading increments shall be applied without impact or jar. The weight of the loading platform or box shall be obtained prior to the test and this weight shall be considered as the first increment of load. If a hydraulic jack is employed, facilities for maintaining each increment of desired load constant under increasing settlement shall be provided. The gauge and the jack shall be calibrated as a unit for each project.

2. The test load shall be twice the proposed working load of the pile. The test load shall be applied in seven increments at a load of 50 per cent, 75 per cent, 100 per cent, 125 per cent, 150 per cent, 175 per cent, and 200 per cent of the proposed working load. After the proposed working load has been applied and for each increment thereafter, the test load shall remain in place until there is no measurable settlement in a 2-hr. period. The total test load shall remain in place until settlement does not exceed 0.001 ft. in 48 hr. The total load shall be removed in decrements not exceeding 25 per cent of the total load at 1-hr. intervals or longer. The rebound shall be recorded after each decrement is removed, and the final rebound shall be recorded 24 hr. after the entire test load has been removed.

3. Under each load increment, settlement observations shall be made and recorded at ½ min., 1 min., 2 min., 4 min., and each 4 min. thereafter after application of load increment, except in the instance of total load where, after the 4-min. reading, the time interval shall be successively doubled until the final settlement limitation is reached and the load is removed.

4. The allowable pile load shall be the lesser of the two values computed as follows:

(a) Fifty (50) per cent of the applied load causing a net settlement of the pile of not more than 0.01 in. per ton of applied load. Net settlement in this paragraph means gross settlement due to the total test load minus the rebound after removing 100 per cent of the test load.

(b) Fifty (50) per cent of the applied load causing a net settlement of the pile of 0.75 in. Net settlement in this paragraph means the gross settlement as defined in 4(a) above, less the amount of elastic shortening in the pile section due to total test load.

(5) FOUNDATION PILES.—Except as provided in section C26-1107.1(c)(1)b6, all building piles within the area of influence of a given load-tested pile of satisfactory performance, shall be installed to the same or greater penetration resistance (or static load) as the successful load-tested pile. The same equipment (or heavier equipment of the same type) that was used to install the load-tested pile shall be used to install all other building piles, and the equipment shall be operated identically. Also all other piles shall be of the same type, shape, external dimension, and equal or greater cross section as the load-tested pile. All building piles within the area of influence represented by a given satisfactory load-tested pile shall bear in, or on the same bearing stratum as the load-tested pile. For friction piles where the actual pile lengths vary more than 50 per cent from that of the test pile, the commissioner may require investigation to determine the adequacy of the piles.

(6) PILE GROUPS.—When the commissioner has reasons to doubt the safe load sustaining capacity of pile groups, he may require at the expense of the owner, group load tests up to 150 per cent of the proposed group load.

(7) "CASING-OFF."—Any temporary supporting capacity that the soil might provide to the pile during a load test, but which would be dissipated with time, shall be obviated by "casing-off" or by other suitable means. For purposes of this section, temporary supporting capacity shall include the resistances offered by any strata of nominally unsatisfactory bearing materials (class 11-65, other than controlled fill) or of soft or loose deposits of class 9-65 or 10-65 that are completely penetrated by the pile, or any resistance offered by granular soils that will be dissipated by reasons of vibration.

(e) Maximum loads.—

(1) BASIC MAXIMUM LOADS.—Except as permitted by the provisions of (2) below, the maximum allowable pile load, determined in accordance with the provisions of sections C26-1107.1(a) through C26-1107.1(d), shall not exceed the values specified in table 11-6.

(2) SUBSTANTIATION OF HIGHER ALLOWABLE LOADS.—The pile capacities tabulated in table 11-6 may be exceeded where a higher value can be substantiated on the basis of test and analysis, as follows:

a. Load tests.—The provisions of section C26-1107.1(d) shall be supplemented as follows:

1. Not less than one single-pile load test shall be conducted for each 10,000 sq. ft. of pile foundation area.

2. Final load increment shall remain in place for a total of not less than 96 hours.

3. Single test piles shall be subjected to cyclical loading or suitably instrumented so that the movements of the pile tip and butt may be independently determined. Other alternate methods or devices, acceptable to the commissioner which will permit evaluation of the transfer of load from piles to soil may be used.

4. Where the commissioner deems necessary, the provisions of paragraph C26-1107.1(d)(6) relating to group load tests shall apply. If required, group load tests shall be performed in groups of numerically average size. Except where the proposed foundation is limited to single and/or two pile groups, each test group shall contain not less than three piles.

5. Individual pile loadings shall not exceed those determined from the single pile load tests.

6. The provisions of paragraph C26-1107.1(d)(2)d shall apply.

b. Analysis and report.—A report shall be submitted by the architect or engineer establishing to the satisfaction of the commissioner (on the basis of soil and load tests and foundation analysis, including analysis of the group action of the piles) that the proposed construction under a 100 per cent overload of the foundation is safe against failure of the pile and soil materials, and showing that the probable total magnitude and distribution of settlement to be expected under design conditions will not result in instability of the building or stresses in the structure in excess of the allowable values established in article 10.

c. Penetration resistance.—The penetration resistance shall not be less than that required by the provisions of section C26-1107.1(c) or, where applicable values are not indicated therein, shall be determined from the required load tests. The pile material shall be capable of withstanding the driving stresses without being damaged.

(f) Combinations of loads.—The provisions of section C26-1101.4 of article 10 shall apply.

§ C26-1107.2 **Allowable lateral load.**—For plumb piles fully embedded in the ground, the lateral load applied at the top of the pile shall not exceed 1 ton per pile unless it has been demonstrated by tests that the pile will resist a lateral load of 200 per cent of the proposed working lateral load without lateral movement of more than 1 in. at the ground level and will resist the proposed working lateral load without a movement of more than ⅜ in. at the ground level. For piles projecting above the ground level the shear and bending stresses computed on the basis of cantilever action to a level of 5 ft. below grade in soils of class 8-65 or better and to 10 ft. below grade in poorer soils shall not exceed the allowable values for like materials established in article 10. The provisions of section C26-1107.1(f) relating to combination of loads shall apply.

§ C26-1107.3 **Uplift capacity.**—A minimum factor of safety against withdrawal of 2.0 shall be provided, except that the factor of safety against withdrawal shall be greater than 2.0 when the piles are subjected to dynamic uplift loads. The uplift capacity shall be demonstrated by pull-out tests, except where a factor of safety of 3 or more based on analysis is used, pull-out tests need not be conducted.

Sub-Article 1108.0 Pile Driving Operations

The provisions of this section shall not apply to piles driven with a vibratory hammer or other equipment wherein the energy of impact cannot be evaluated.

§ C26-1108.1 **Equipment.**—

(a) General.—The hammer shall travel freely in the leads. The cushion or cap block shall be replaced, if crushed. The hammer shall deliver its rated energy, and measurements shall be made of the fall of the ram or other suitable data shall be observed at intervals as required to verify the actual energy delivered at the termination of driving.

(b) Cushion or cap block.—The cushion or cap block shall be a solid block of hardwood with its grains parallel to the axis of the pile and enclosed in a tight-fitting steel housing, or shall be an equivalent assemblage. If laminated materials are used, the type and construction of these materials shall be such that their strength is equal to, or greater than, hardwood. Wood chips, pieces of rope, hose, shavings, or automobile tires and similar materials shall not be used. Cap blocks shall be replaced if burned, crushed, or otherwise damaged.

(c) Followers.—Followers shall not be used unless permitted in writing by the architect or engineer, and only when necessary to accomplish such installation. They shall be of steel or hardwood of such size, shape, length, and weight as to permit driving the pile in the desired location and to the required depth and resistance. The required driving resistance tabulated in Tables 11-4 and 11-5 shall be increased to compensate for the loss of energy in the hammer blow. The follower shall be a single length section, shall be provided with a socket or hood carefully fitted to the top of the pile to minimize loss of energy and to prevent damage to the pile, and shall have sufficient rigidity to prevent "whip" during driving.

§ C26-1108.2 **Procedures.—**

(a) Continuous Driving.—Driving of piles preliminary to final seating shall be continuous for an interval adequate to break or prevent the development of freeze. The hammer shall be operated at its rated speed during this interval.

(b) Jetting.—Jetting, augering and other methods of pre-excavation shall not be used unless permitted in writing by the architect or engineer. When permitted, such procedures shall be carried out in a manner which will not impair the carrying capacity of the piles already in place or the safety of existing adjacent structures. Jetting or augering shall be stopped at least 3 ft. above the final expected pile-tip elevation and at least 3 ft. above the tip elevation of any pile previously driven within 6 ft. of the jet or auger, except that where piles are to be end bearing on rock or hardpan, jetting or augering may be carried to the surface of the rock or hardpan. Piles shall be carried down beyond the depth of jetting, augering or other pre-excavation until the required resistance is obtained. If there is evidence that jetting or other procedures have disturbed the load-bearing capacities of previously installed piles, those piles that have been disturbed shall be restored to conditions meeting the requirements of this article by redriving or by other methods acceptable to the architect or engineer. Redriving or other remedial measures shall be instituted after the jetting or other operations in the area have been completed.

(c) Sequence of installation.—Individual piles and pile groups shall be installed in such sequence that: (1) the carrying capacity of previously installed piles is not reduced; (2) the soil surrounding the piles is not compacted to the extent that other piles in the group cannot be installed properly; and (3) ground movement that would damage adjacent buildings or utilities is prevented. In general, in any group, driving shall start from one side and proceed progressively toward the other side of the group or shall start from the middle and proceed toward the edges of the group.

(d) Heaved piles.—In soils in which the installation of piles causes previously installed piles to heave, accurate level marks shall be put on all piles immediately after installation, and all piles that have been heaved by an amount in excess of ¼ in. shall be redriven to the required resistance.

(e) Penetration measurements.—Penetration measurements made for the purpose of determining resistance to driving shall not be made when pile heads are damaged to an extent that may affect measured penetration, nor shall they be made immediately after fresh cushion blocks have been inserted under the striking part of the hammer.

Sub-Article 1109.0 Pile Types—Specific Requirements

§ C26-1109.1 **Scope.**—Types of pile construction and installation that are not described in this section will be permitted for use only where acceptable to the commissioner.

§ C26-1109.2 **Timber piles.**—

(a) Materials—Timber piles shall conform in quality to class A or B of Reference Standard RS 11-7.

(1) SIZE OF PILES.—Piles shall be of adequate size to resist the applied loads without creating stresses in the pile materials in excess of 1200 psi for piles of southern pine, Douglas fir, oak, or other wood of comparable strength; or 850 psi for piles of cedar, Norway pine, spruce or other wood of comparable strength. Except as provided in (2) below, for piles 40 ft. or more in length and of 30 tons capacity or less, the following sizes or classes of piles shall be deemed to be adequate for considerations of stress in the pile material:

a. Piles of 25 to 30 tons capacity—Class A or minimum 8 in. tip with uniform taper.

b. Piles of less than 25 tons capacity—Class A or B or minimum 6 in. tip with uniform taper.

(2) PILES DRIVEN TO END BEARING.—All timber piles, regardless of capacity, driven to end bearing on soils of classes 1-65 through 5-65 shall be class A or shall have a minimum 8 in. tip and a uniform taper.

(3) SPECIES OF WOOD.—Any species of wood may be used that conforms to the provisions of reference standard RS 11-7 and that will stand the driving stresses.

(4) PEELING.—Unless treated, piles need not be peeled.

(b) Limitations on use.—Where timber piles are to be driven to end bearing on soils of classes 1-65 through 5-65 by use of an impact hammer, the installation of each such pile shall be under the personal supervision of an architect or engineer, and the operations of driving such piles, the observations of penetration resistance, and the operation of the equipment shall be so conducted as to terminate driving directly when the pile reaches bearing on the hard material. A report, prepared by the architect or engineer, describing the procedures, equipment, and precautions followed to prevent injury to the piling shall be submitted to the commissioner.

(c) Lagged and inverted piles.—The use of lagged or inverted piles will be permitted. Double lagging shall be adequately connected to the basic pile material to transfer the full pile load from the basic pile material to the lagging without exceeding values of allowable stress as established in article 10. The connection for single lagging shall be proportioned for half the pile load. The diameter of any inverted timber pile at any section shall be adequate to resist the applied load without exceeding the stresses indicated in (a) (1) above, but in not case shall it be less than 8 in.

(d) Installation.—

(1) All broomed, crushed, or otherwise damaged materials at the head of the pile shall be removed before capping.

(2) Any sudden decrease in driving resistance shall be investigated with regard to the possibility of breakage of the pile, and if such sudden decrease in driving resistance cannot be correlated to boring data or some incident in the driving, and if the pile cannot be removed for inspection, it shall be considered as adequate cause for rejection of the pile.

§ C26-1109.3 **Precast concrete piles (including prestressed sections).**

(a) Materials.—Materials for precast concrete piles shall conform to the requirements of section C26-1004.0.

(b) Construction.—

(1) HANDLING.—Precast concrete piles shall be proportioned, cast, cured, handled, and driven so as to resist the stresses induced by handling and driving as well as by loads. Handling stresses shall be computed on the basis of 50 per cent of the weight of the pile as an allowance for impact. Handling equipment shall be constructed so as to equalize the reactions on the several lines of the pile pickups. Loading conditions induced by handling and driving shall be considered as of infrequent occurrence.

(2) DIMENSIONS.—The minimum lateral dimension of the pile shall be 10 in., except for the taper at the tip.

(3) STRUCTURAL DESIGN.—Piles shall be proportioned in accordance with the provisions of sub-article 1107.0, subject to the following additional requirements:

a. For a length equal to at least three times the minimum lateral dimension at each end of the pile, lateral tie reinforcement consisting of 0.225 in. diameter rods or larger shall be spaced not more than 3 in. center-to-center, or an equivalent spiral shall be provided. Elsewhere, the spacing of the ties or the pitch of the spiral may be increased to 12 in. The minimum amount of longitudinal reinforcement shall be 2 per cent of the concrete section placed in a symmetrical pattern of at least four bars. If prestressed piles are used, the minimum residual compression in the pile section shall be 700 psi. For piles designed with voids, the 3 in. spacing of the ties or spiral at each end of the pile shall be extended for a distance of 12 ft. or ⅓ the length of the pile, whichever is smaller.

b. For piles designed with voids, the minimum wall thickness of the concrete in any section of the piles shall be 4 in. Voids may extend through either or both ends of the pile. If the void extends through the lower end of the pile, the pile head shall be vented to prevent buildup of internal hydraulic pressure during driving.

c. Reinforcing steel shall be covered with at least 1½ in. of concrete on the surface against the ground.

(c) Tolerances.—Voids, when used, shall be located within ⅜ in. of the position shown in the plans. The maximum departure of the pile axis from a straight line, measured while the pile is not subject to bending forces, shall not exceed ⅛ in. in 10 ft. or ⅜ in. in 40 ft., or 0.1 per cent of the pile length.

(d) Installation.—Precast concrete piles shall not be handled or driven until they have cured sufficiently to develop the necessary strength.

§ C26-1109.4 **Cast-in-place concrete piles.—**

a) Description.—Cast-in-place concrete piles shall be cast in shells previously installed in the ground or, with the limitations indicated in section C26-1106.10, may be cast in an uncased hole. Cast-in-place piles may be tapered or cylindrical, or a combination of tapered and cylindrical shapes.

b) Materials.—Concrete for cast-in-place concrete piles shall conform to the requirements of sub-article C26-1004.0. Slump shall be 5 in. plus or minus 1 in. Where the pile shell is assumed to contribute to the capacity of the pile as a structural member, the shell shall be of metal and shall conform to the provisions of sub-article C26-1005.0 or C26-1109.7. Where the pile shell is not assumed to contribute to the capacity of the pile as a structural member, the shell may be of any material that will adequately resist the driving stresses and maintain an open well to receive the concrete. End closures for shells shall not project more than ½ in. beyond the outer limits of the pile when bearing in soils of class 6-65 or lower.

c) Installation.—

(1) After installation to final depth and immediately before filling with concrete, the inside of the tube, shell, or bore shall be thoroughly cleaned to the bottom and inspected by lowering a drop light or by means of a light beam. To be accepted: (a) the pile shall be free of collapsed sections of shell and the pile shell shall not show any tears; (b) the pile shall be free of water except that a minor amount of water may be allowed to occur in the pile if it be absorbed by placing a suitable amount of dry cement-sand mixture in the tip end of the pile; and (c) the alignment of the pile shall conform to the provisions of section C26-1106.4. If the bottom of the casing is out of sight, the shape and alignment of the casing shall be surveyed with a suitable instrument, or the pile rejected.

(2) Concrete shall be placed by such methods that the entire volume of the tube, shell, or bore is filled and in a manner that will preclude separation of the ingredients.

(3) No concrete shall be placed in a cast-in-place pile until all piles within a radius of 15 ft., or within the heave range, have been driven.

(4) Rejected pile shells shall be filled with concrete or sand.

5) The concrete cap shall not be placed until at least 1 hr. after all piles within the cap group are completely filled.

§ C26-1109.5 **Compacted concrete piles.—**

(a) Description.—A "compacted concrete pile" shall denote a concrete pile formed with an enlarged base in which the concrete in the base is placed in small batches that are compacted prior to attaining an initial set. The concrete in the shaft of the pile shall be placed as specified in section C26-1109.4 for cast-in-place concrete piles if a permanent casting is provided, or in small batches that are compacted.

(b) Materials.—Concrete for compacted concrete piles shall conform to the requirements of sub-article C26-1004.0. Concrete to be compacted shall have a minimum compressive strength at the age of 28 days of 4000 psi and shall be mixed with sufficient water to permit hydration of the cement, but the slump shall be zero. The concrete shall be placed and compacted before initial set can occur. Non-compacted concrete, if used for the pile shafts, shall conform to the requirements for cast-in-place concrete piles.

(c) Spacing.—Minimum spacing between compacted concrete piles shall be 4 ft.-6 in., center to center except that where the shafts of such piles are cased for their full length, this spacing may be reduced to 3 ft.-6 in. Where a question exists as to possible damage to adjacent previously driven piles, these minimums shall be increased.

(d) Installation.—

(1) The base shall be formed by ramming concrete, in batches of approximately 5 cu. ft. or less, from a drive casing and into the soil. Unless specifically otherwise permitted by the commissioner, a minimum of 20 blows of at least 140,000 ft.-lbs. per blow shall be required for extrusion of the last 5 cu. ft. of concrete. The total quantity of concrete extruded from the drive casing to form the base shall be equal to or greater than the quantity so extruded in the case of the nearest successful applicable test pile, except that a compactive effort in excess of 30 blows, each of 140,000 ft.-lbs., will not be required for extrusion of the last 5 cu. ft.

(2) After the expanded base has been formed, the shaft shall be constructed. Where a cased shaft is to be used, a steel shell shall be inserted into the drive casing and anchored to the expanded base by placing a fresh charge of concrete in the shell and driving it into the base. The shell may then be filled with concrete to cut-off elevation after the removal of the drive casing, in accordance with the provisions relating to cast-in-place concrete piles. Any annular space remaining between the shell and sur-

rounding soil may be suitably filled to assure proper lateral support of the shaft, unless there is sufficient recovery of the ground to provide the necessary support. Where an uncased shaft is to be used, the provisions of section C26-1106.10 shall apply, and the concrete for the shaft shall be placed by ramming or by the use of approved pressure devices as the drive casing is withdrawn. Where ramming is used, the concrete batches being rammed shall not exceed 5 cu. ft. in volume, not less than two blows of 30,000 ft.-lb. each shall be applied to compact each batch of concrete, and the bottom of the drive casing shall be kept below the level of the concrete at all times.

(3) The outside diameter of the permanent shaft shall not be more than 4 in. less than the inside diameter of the drive casing.

(4) Except where uncased shafts are used, as described in section C26-1106.10, no concrete shall be placed in the pile shafts until all piles within a radius of 15 ft., or within the heave range, have been driven.

(e) Bearing material.—The enlarged base of the pile shall be formed in, or on the same type of bearing material as is used to support the nearest applicable load test pile and at a similar depth therein. In addition, the enlarged base shall be underlain by a minimum depth of 10 ft. (measured from the junction of the shaft and base) of soil materials of classes 1-65 to 7-65, except that, where installation of the base is permitted to be performed with blows of less energy than the 140,000 ft.-lbs. indicated in section C26-1109.5(d)(1), the requirement for a 10 ft. depth of class 1-65 to 7-65 material may be reduced, subject to the approval of the architect or engineer and the approval of the commissioner.

§ C26-1109.6 **Steel H sections.—**

(a) Materials.—Steel H sections may be of any type of steel permitted by the provisions of reference standard RS 10-5. The use of built-up sections or sections of other than "H" form will be permitted if the several components of the section are adequately connected to develop the strength of the adjacent components and if the ratio of width to thickness of the component parts does not exceed the values for conventional "H" sections.

(b) Limitations on use.—The tips of all steel H piles having a thickness of metal less than 0.5 in., which are driven to end bearing on rock of class 1-65 through 3-65 by an impact hammer, shall be reinforced. The installation of all steel H piles by impact hammer to end bearing on rock of classes 1-65 through 3-65 shall be under the personal inspection of an architect or engineer, and the operations of driving such piles, the observations of penetration resistance, and the operation of the equipment shall be conducted so as to terminate driving directly when the pile reaches refusal on the rock surface.

§ C26-1109.7 **Concrete-filled pipe piles.—**

(a) Materials.—The pipe shall conform to the provisions of reference standard RS 11-8. Concrete shall conform to the requirements of subarticle C26-1004.0.

(b) Minimum dimensions.—

(1) Pipe installed open-end and having a nominal outside diameter of less than 14 in. shall be at least 0.25 in. thick. For diameters from 14 in. to 18 in., the minimum thickness shall be 0.310 in. For diameters over 18 in., the minimum thickness shall be 0.375 in.

(2) Steel pipe piles installed with ends closed shall have a minimum nominal wall thickness of at least 0.125 in.

(3) For piles which receive their principal support by friction, closure or splice plates shall not project more than ⅜ in. beyond the outer limit of the pipe section.

(c) Installation.—

(1) Pipe shells driven open-end shall be cleaned to the bottom of the shell after driving.

(2) After driving and cleaning the pipe, open-end piles driven to end bearing on rock or hardpan shall be reseated to full bearing by redriving, to the resistance indicated in Table 11-4. If the pipe shell shows 2 in. or more of penetration on redriving, the pipe shall be recleaned and redriven in successive cycles until the penetration on redriving is less than 2 in.

(3) Pipe shells shall be inspected before filling with concrete, shall be clean, and shall meet the requirements for alignment and condition of the shell as specified with regard to the shells of cast-in-place piles. If leakage of water into the pipe occurs, the provisions of section C26-1109.8(e)(5) shall apply.

(4) Placing of concrete fill in pipe shells shall conform to the requirements for placing concrete fill in cast-in-place piles.

§ C26-1109.8 **Caisson piles.—**

(a) Description.—Caisson piles shall denote concrete filled pipe piles that are socketed into bedrocks of class 1-65, 2-65 or 3-65 and constructed with steel cores.

(b) Materials.—Pipe or shell and concrete shall conform to the requirements for concrete filled pipe piles, except that the minimum compressive strength of the concrete at the age of 28 days shall be 3,500 psi. Steel cores shall conform to the requirements for steel H piles. Reinforcing steel cages shall be covered with at least 1½ in. of concrete.

(c) Design of rock socket.—The design of the rock socket shall be predicated on the sum of the allowable bearing pressure on the bottom of the socket plus bond along the sides of the socket. The allowable bearing pressure on the surface of the rock at the bottom of the socket shall be as established in section C26-1103.4 increased for embedment in accordance with note 8 of table 11-2, provided that the strength of the concrete fill in the socket, computed as $0.45f'^{c}$ is of comparable magnitude. The allowable bond stress between the concrete and the sides of the socket shall be taken as 200 psi. The provisions of section C26-1107.1(c) relative to penetration resistance shall not apply.

(d) Spacing and minimum dimensions.—

(1) Minimum diameter of a caisson shall be 18 in. with a minimum shell thickness of ⅜ in. Minimum depth of the rock socket shall be equal to the diameter of the pipe.

(2) The center-to-center spacing of caissons shall be at least 2½ times the outside diameter of the shell.

(e) Installation.—

(1) The steel shell shall be installed through overburden the material within the shell shall be removed, and the shell seated in the rock sufficiently to stop the inflow of soil. Where required to extend the shells, splices are to be welded. A suitable steel driving shoe shall be welded to the bottom of each caisson.

(2) A socket shall then be drilled in the rock to the required depth and shall be approximately of the same diameter as the inside diameter of the shell. Before placement of concrete, the socket and shell shall be thoroughly cleaned and the rock inspected to verify that the rock is of the class on which the design has been predicated, or of a better class. In case visual inspection cannot be made because of inability to unwater the caissons by standard pumping methods, drilling logs and screenings from the rock drilling operation may be utilized to determine the class of rock in the socket.

(3) Where more than one section of steel core is required, the mating ends of the sections shall be spliced so as to safely withstand the handling stresses to which they may be subjected. The ends shall be milled or field ground to insure contact. The steel core shall be centrally installed in the caisson before grouting and concreting, shall not be more than 1 in. above the rock at the bottom of the socket, and shall be full length of the caisson or extend a sufficient distance up into the shaft to transmit the load in the steel core into the concrete of the caisson. A minimum-weight 36 lb. stub core beam shall be installed in the socket for caissons not requiring steel cores in order to lock the caissons into the rock. In these cases, the length of the steel cores shall be twice the socket depth.

(4) Concrete and grout shall be placed so that it completely fills the shell, the socket, and the space between the steel core and shell, and in a manner that will preclude separation of the ingredients.

(5) If the leakage of water into the caisson is minor, the caisson shall

be pumped out and 1 cu. yard of grout shall be placed in the caissons and then the balance of the concrete installed. If the leakage of water makes it inadvisable to attempt to place concrete in the dry, the shell shall be filled to its top with clean water, and the concrete placed by the tremie method to the top of the caisson in one continuous operation or by using a seal of grout of the same strength as the specified concrete. The grout seal, if used, shall be deposited by means of a grout pipe to an elevation of at least 3 ft. above the cutting edge, and after a sufficient time has elapsed to allow the grout to set, the caisson shall be pumped dry and the remaining space filled with concrete.

§ C26-1109.9 **Composite piles.**—Composite piles include those consisting of two types of pile joined together. The maximum allowable load shall be that allowed for the component of lesser strength used to make up the full pile length. The connection or joint between the two components shall be constructed so as to prevent the separation of the upper and lower components during construction and thereafter. The details and methods of making joints shall be designed.

Sub-Article 1110.0 Underpinning

§ C26-1110.1 **General requirements.**—Where support of adjacent structures or properties is required, such support may be provided by underpinning, sheeting, and bracing, or by other means acceptable to the commissioner. Except as specifically permitted otherwise, underpinning piers, walls, piles, and footings shall be designed and installed in accordance with the applicable provisions of this article relating to piers, walls, piles, and footings used in new construction and shall be inspected as provided in section C26-1112.6.

§ C26-1110.2 **Use of rock support in lieu of underpinning.**—Existing structures founded at a level above the level of adjacent new construction may be supported on hard rock in lieu of underpinning, the use of sheeting and bracing, or the construction of retaining walls, provided that a report by the architect or engineer is submitted substantiating the safety of the proposed construction and verifying that an "in-place" inspection has been made of the rock exposed and of the jointing therein in the excavation.

Sub-Article 1111.0 Stability

§ C26-1111.1 **General.**—The possibility of overturning and sliding of the building shall be considered.

§ C26-1111.2 **Factor of safety.—**

(a) Overturning.—The minimum factor of safety against overturning of the structure as a whole shall be 1½. Stability against overturning shall be provided by the dead load of the building by the allowable uplift capacity of piling, by anchors, by the weight of soil directly overlying footings provided that such soil cannot be excavated without recourse to major modification of the building, or by any combination of these factors.

(b) Sliding.—The minimum factor of safety against sliding of the structure under lateral load shall be 1½. Resistance to lateral loads shall be provided by friction between the foundation and the underlying soil, by passive earth pressure, by batter piles, or by plumb piles, subject to the following:

(1) The resistance to lateral loads due to passive earth pressure shall be discounted where the abutting soil could be removed, inadvertently, by excavation.

(2) In the case of pile supported structures, frictional resistance between the foundation and the underlying soil shall be discounted.

(3) The available resistance to friction between the foundation and the underlying soil shall be predicated on an assumed friction factor of 0.5 for soils of classes 1-65 through 8-65. A greater value of coefficient of friction may be used subject to verification by analysis and test. For soils of poorer classes, the stability shall be analyzed by accepted procedures of soil mechanics.

Sub-Article 1112.0 Inspection

§ C26-1112.1 **General.**—The applicable provisions of sub-article C26-106.0 shall apply.

§ C26-1112.2 **Boring operations.**—Boring operations shall be subject to controlled inspection, except that 50% or less of the required number of borings may be inspected by an architect or engineer other than the architect or engineer designated for controlled inspection. The records of borings shall be attested to as follows:

(a) The architect or engineer shall file a report stating which borings were performed under his inspection and whether such inspection was performed personally or otherwise. If the inspection was not made personally by the architect or engineer, the name and address of the inspector shall be noted. It shall be stated: that the borings so inspected were made and were carried to the depths indicated; that, to the best of the architect's or engineer's knowledge and belief, the description and classification of the soils are a true description of the samples recovered from the respective borings; that such samples were recovered at the levels indicated; and that the boring work progressed in such manner that the samples recovered are reasonably representative of the subsurface conditions.

(b) The accuracy of the other data indicated on the boring records shall be attested to by the drilling contractor or by the driller making the borings.

§ C26-1112.3 **Piling.**—The installation of all piling shall be subject to controlled inspection. Such inspection shall be performed only by an architect or engineer resident at the site, except that where more than one pile rig is working at a site, inspection of the work may be performed by nonlicensed or nonregistered personnel working under the resident architect or engineer, who need act only in a supervisory capacity. This exception shall not apply, however, in cases of timber or steel piles driven to end bearing as described in sub-article C26-1109.0. Materials for piling shall be subject to inspection requirements as prescribed in article 10 for those or like materials. In all cases, an inspector shall be assigned to observe the operations of each rig.

§ C26-1112.4 **Footings, foundation piers, foundation walls and pile caps.**—The provisions of section C26-1105.1 shall apply.

§ C26-1112.5 **Subgrade for footings, foundation piers, and foundation walls.**—The soil material directly underlying footings, foundation piers, and foundation walls shall be inspected by an architect or engineer after excavation and immediately prior to construction of the footings. If such inspection indicates that the soil conditions do not conform to those assumed for purposes of design and described on the plans, or are unsatisfactory due to disturbance, then additional excavation, reduction in allowable bearing pressure, or other remedial measures shall be adopted, as required. A copy of a report or reports on such inspection or inspections describing the conditions found and any necessary modification of the design, and bearing the signature of the architect or engineer making the inspections, shall be filed with the commissioner. In addition, notification shall be received by the department at least two working days prior to construction of the footing, pier, or foundation walls, that the subgrade is ready for inspection unless the requirements of section C26-120.5 have been met.

§ C26-1112.6 **Construction required for or affecting the support of adjacent properties or buildings.**—Except in cases where a proposed excavation will extend less than 10 ft. below the legally established grade, all underpinning operations and the construction and excavation of temporary or permanent cofferdams, caissons, braced excavated surfaces, or other constructions or excavations required for or affecting the support of adjacent properties or buildings shall be subject to controlled inspection. The details of underpinning, cofferdams, caissons, bracing, or other constructions required for the support of adjacent properties or buildings shall be shown on the plans or prepared in the form of shop or detail drawings and shall be approved by the architect or engineer who prepared the plans.

PART II

Identification of Minerals and Rocks

Rocks are made up of minerals which may be recognized by skillful geologists. The competent engineer, working with foundations in rock, should be prepared to recognize general categories of minerals and the properties that may affect a given project.

The following table lists some of the common minerals and their properties:

Table A.1

Name	Luster	Hardness (*Moh*)	Specific Gravity	Other Properties
Quartz (SiO2)	Vitreous to greasy	7	2.65	Commonly translucent to opaque
Orthoclause KALSi308	Pearly to vitreous	6	2.5	Pink or gray
Serpentine Mg3Si205(OH)	Greasy	3	2.6	Green—smooth greasy feel
Calcite CaCO3	Vitreous to dull	3	2.7	Effervesces with dilute HCl
Gypsum CaSO4*2H20	Vitreous to pearly	2	2.3	Usually white or colorless
Halite NaCl	Vitreous	2.5	2.1	Strong salty taste
Magnetite Fe304	Metallic	6	5	Black and attracted by magnet
Pyrite FeS2	Metallic	6	5	Brassy yellow
Galena PbS	Metallic	2.5	7.5	Silver gray to black
Uraninite UO2	Dull metallic	5	7.5	Gray or black

The major classes of rocks consists of combinations of minerals. There are approximately 30 kinds of rocks that make up the earth's crust. An engineer interested in rock work should be able to recognize the more common varieties and be knowledgeable of their properties.

The igneous rocks are the most systematic of the three classes and are grouped conveniently according to texture and composition. All rocks in the same horizontal column are alike in texture; those in the same vertical

Table A.2 Igneous Rock Composition

Texture	Silica Content			
	High			Low
Granular grains nearly equal size	Granite	Diorite	Gabbro	Peridotite
Grains visible with magnifying glass (Aphanitic)	Rhyolite	Andesite	Basalt	—
Glassy	Pumice	Basalt glass	—	—
Pyroclastic	Volcanic tuff		—	—

column are generally alike in chemical composition. Light-colored igneous minerals have a high content of silica, the dark colored have less. The coarse grained rocks, such as granite, are formed deep in the crust, where slow cooling favors crystal growth; the fine granular and porphyritic rocks are most abundant at shallow depths; and the glassy rocks are volcanic rocks chilled quickly in the earth's surface.

The same magma that produces granite with slow cooling deep in the crust produces rhyolite with rapid cooling in a lava flow. Exceptionally rapid cooling produces glassy lavas, and volcanic tuff is produced by explosive eruptions. Granite is made up of feldspar, quartz, and miscellaneous minerals and is widespread on all the continents. The miscellaneous minerals commonly are in almost perfect crystals, which suggests that they formed earliest, while most of the mass was molten. The feldspar formed next, and the grains crowded against each other in growth. Quartz, the surplus silica, crystallized last and therefore is molded around the angular grains of the earlier minerals. This interlocking arrangement of grains is characteristic of granite and is called its texture.

Diorite is made up of feldspar mixed with impurities. Generally the impurities are more abundant than in granite. Diorite forms many large masses, but is not as common as granite.

Gabbro—there is a gradation from dark diorite into gabbro as the feldspar content becomes subordinate to the other minerals. These minerals (usually pyroxene) are heavier than feldspar and gabbro is distinctly heavier than granite or diorite.

Peridotite is a granular rock made up of minerals such as pyroxene and olivine and contains little feldspar. These rocks are dark and heavy, and are commonly associated with ores containing the metals, nickel, platinum, and iron. In many masses the pyroxene and olivene have been altered to serpentine, probably by hot water and gases.

Rhyolite is a fine-grained rock with some large crystals of quartz or feldspar, which indicates an excess of silica. Colors of the ground mass range from almost white to shades of gray, yellow, red, or purple. It commonly has irregular bands made by flowage of the stiff magma.

Andesite is a fine-grained rock generally similar to rhyolite but lacking the quartz crystals. It usually contains crystals of feldspar and dark minerals. Common colors are gray and green, but it may be black. Andesite is abundant as a volcanic rock around the margins of the Pacific.

Basalt is a fine-grained rock that is opaque (which distinguishes it from andesite). Common colors are black and dark gray. In the upper part of lava flows it may be filled with small holes made by escaping gases; it is the most abundant rock in lava flows.

Glassy rocks are formed by magmas that chill rather quickly. Obsidian is the highly lustrous kind and pitchstone has a dull greasy luster. Natural glass forms crusts on lava flows of rhyolite and andesite.

Pumice is glass froth full of spaces made by gases. For this reason its specific gravity is usually less than that of water.

Basalt glass forms as a thin coating at the tops of some baltic flows. It is opaque, like ordinary basalt, but has a glassy luster.

Pyroclastic rocks are formed by tiny fragments of rock that have been blown over the landscape during an eruption. The particles of ash are very hot and when spread thickly the temperature is high enough to cause partial remelting. The ash particles are thus welded together and the resulting rock has a glassy appearance.

Many volcanic tuffs are bedded and look much like sedimentary rocks. These rocks are thus special in classification: partly igneous and partly sedimentary.

The sedimentary rocks are usually divided into two kinds of divisions: (a) the clastic rocks made of particles derived by weathering and erosion of older rocks; and (b) the chemical and organic rocks made of minerals either precipitated chemically or accumulated by the activities of plants and animals.

The clastic rocks are classified according to the size of grain before consolidation.

Conglomerate or breccia is a gravel or broken rock that has been cemented together with finer grained particles of sand mixed with silt or clay. The binder usually consists of silica iron oxide or calcium carbonate.

Sandstone is made of sand grains cemented with calcium carbonate or silica.

Shale is made of compacted clay and silt. It generally has a smooth or greasy feel. Many shales split into thin layers or flake. It is the most abundant of the sedimentary rocks.

Chemical and organic rocks are classified according to the process that preceded their structuring.

Limestone consists chiefly of calcite and is formed by the precipitation and consolidation of shells. It effervesces freely in hydrochloric acid.

Chalk is weakly cemented limestone usually containing many very small shells. Marl is a mixture of calcium carbonate and clay.

Dolomite is slightly harder than limestone and effervesces with acid when scratched or powdered. The rock is made chiefly of the mineral dolomite, which has magnesium and calcium in its composition.

Peat and coal are formed by the consolidation of decomposed plant tissues. Lignite or brown coal is much more compact than peat. Bituminous or soft coal is black and breaks into blocks. Anthracite or hard coal breaks with conchoidal fracture and has a high luster. Anthracite represents an early stage in metamorphism, but it is included in the sedimentary rock classification.

Salt and gypsum occur in large deposits, and in the pure form they have distinctive properties, such as taste and hardness.

Metamorphic rocks are caused by crustal deformation and by igneous activity. Many metamorphic rocks are foliated; that is, they consist of minerals such as micas and chlorite that are set in parallel arrangement.

Slate has a plane cleavage that cuts across the bedding planes of the original sedimentary rock. Thin slabs ring when they are tapped sharply. Common colors are gray, red, green, and black.

Phyllite is a lustrous rock representing a later stage of metamorphism than slate. The mica flakes responsible can be seen only with a microscope. Phyllite grades from slate to schist.

Schist is classified as a well-foliated rock in which the flaky minerals are visible. Mica schist is found extensively throughout the New York City area. Quartz is abundant in all kinds of schists.

Gneiss is a coarse-grained rock with imperfect foliation. Many gneisses are streaky and roughly layered, which is caused by alternating lenses that differ in composition. Feldspar, quartz, mica, hornblande, and garnet are minerals commonly found in gneiss formations.

Marble is crystalline limestone or dolomite. Impurities in the original rock form striking patterns in marble formations.

Quartzite is a rock that has quartz grains fully cemented with silica. The cementing material has the same composition and physical properties as the original grains. Quartzite fractures pass through the grains of the original aggregate and not around them as in a sandstone.

References

1. C. R. Longwell and A. F. Flint, *Introduction to Physical Geology,* 2nd ed., Wiley, New York, 1962.
2. C. D. Dunbar and K. M. Waage, *Historical Geology,* 3rd ed., Wiley, New York, 1969.

Index